AF386391

In Company
with Tanks

This book is dedicated to the riflemen of G Company, 7 Rifles,
in memory of their predecessors from east London and
further afield who fought with 8th Rifle Brigade in Normandy.

In Company with Tanks

8th Rifle Brigade in Normandy

Tim Saunders

Pen & Sword
MILITARY

First published in Great Britain in 2026 by
PEN & SWORD MILITARY
an imprint of Pen & Sword Books Ltd
Yorkshire – Philadelphia

ISBN 978-1-03613-592-8

Typeset by Concept, Huddersfield, West Yorkshire, HD4 5JL.
Printed and bound in England by CPI Group (UK) Ltd, Croydon, CR0 4YY.

The Publisher's authorised representative in the EU for product safety is Authorised Rep Compliance Ltd, Ground Floor, 71 Lower Baggot Street, Dublin D02 P593, Ireland – www.arccompliance.com

For a complete list of Pen & Sword titles please contact
PEN & SWORD BOOKS LTD
47 Church Street, Barnsley, South Yorkshire, S70 2AS, England
E-mail: enquiries@pen-and-sword.co.uk
Website: www.pen-and-sword.co.uk
or
PEN & SWORD BOOKS
1950 Lawrence Rd, Havertown, PA 19083, USA
E-mail: uspen-and-sword@casematepublishers.com
Website: www.penandswordbooks.com

Contents

Acknowledgements

This book is principally based on the work of four of 8th Battalion, The Rifle Brigade's, officers, who produced the booklets covering their company's experiences in the North-West European Campaign. They were helped by other soldiers who added their memories of battles that had so recently finished. Since the immediate aftermath of the war, a number of members of 8 RB have also written or given accounts that are to be found in print and online. Writing during the battles recounted here were a succession of adjutants and intelligence officers who maintained the battalion's war diary, which has also been an essential base document.

Two of the company booklets are readily available but for the other two, I am especially grateful to the curator and volunteers of the Green Jackets and Rifles Museum in Winchester for essential access and help in copying the booklets, war diaries and documents. In this respect I must also thank Lieutenant Colonel Robin Davies and as ever my co-conspirator Richard Hone, whose wealth of technical knowledge of the Second World War I could not have done without. My thanks are due to Andrew Newson, who runs the British Army War Diary Copying Service and has saved me considerable amounts of time and money with his help.

As ever I have been encouraged and assisted by friends and colleagues with historical advice and the battlefield guides' knowledge of the ground in Normandy.

Finally, as ever I am indebted to the services of Pen & Sword team namely, Heather, Paul and the incomparable designer Noel. Once again, I owe them my grateful thanks for their patience in nursing this project to fruition.

Thank you one and all.

Tim Saunders MBE
Warminster, 2026

Tim Saunders is an accredited member (006)
of the International Guild of Battlefield Guides.

Introduction

For the four companies of 8th Battalion, The Rifle Brigade to each write their own short history is almost unique.[1] The battalion rarely fought as a whole, with its companies normally grouping with 'their' particular armoured regiment of 29 Armoured Brigade and, therefore, they had their own stories to tell. The resulting booklets, however, were aimed as mementos for those who survived the campaign and as memorials to the men who numbered amongst the battalion's killed. As these booklets were written in house for an internal audience, they leave out much of the wider context and detail of daily routine that would have been well known to those who fought with 8 RB. This book, therefore, aims to brings together the stories of the men who fought in the British Army's premier armoured division fighting in the Normandy Campaign for modern readers.

While the emphasis is, of course, on the story of 8 RB, as the motor companies fought for much of the time with the armoured regiment, there is a need to include much of their activity to produce a more complete story.

In many respects the title of 'motor battalion' was a misnomer, but one that was, of course, correct when infantry battalions in armoured brigades were equipped with 15-cwt trucks and White scout cars. With, however, the issue of the half-track, they like their US Army counterparts became 'armoured', all be it only lightly so. As the 8th Battalion's replacement vehicles arrived within weeks of D Day, it is easy to understand that changing nomenclature at that stage was the least of the army's concerns, so the misnomer endured.

The motor battalions were at the forefront of developing battlegroups and associated tactics across the three British armoured divisions in Normandy, shrugging off outdated doctrine and desert experience in the process. In the bocage country tanks and motor companies learned from 7th Armoured Division's experience during Operation PERCH, and closer to home 29 Armoured Brigade learned valuable lessons in both operations EPSOM and GOOD-WOOD. By the end of the Normandy Campaign, 11th Armoured Division under command of Major General Pip Roberts was both flexible and effective, no small part due to 8 RB.

A couple of notes: Firstly, to make a sensible sequence of events, the times quoted in German sources have been converted to double British summer time as used by the Allies during the Normandy Campaign. Secondly, as is my usual practice I have used the ranks whose fighting is covered in this book, as they were at the time. Many were, of course, promoted to higher ranks later in the campaign or post-war.

A recruitment poster issued during the doubling of the TA in 1938/39.

Glossary of Terms

All wars produce their own lexicon. Therefore, it is no surprise that abbreviations abound in war diaries and accounts of the North-West Europe Campaign. The main ones are listed here, but in an army where combat training had priority over staff work, many non-standard and local variations were used. The following list will help:

2iC	Second in Command
A1	Echelon Immediate replenishment of combat supplies
A2	Echelon Further back planned replenishment of combat supplies
AA	Anti-aircraft
ACV	Armoured Command Vehicle
ADC	Aide de Camp
Adjt	Adjutant, unit commanding officer's staff officer
Admin	Administration
Adv	Advance
AGRA	Army Group Royal Artillery, a corps' artillery 'brigade', consisting of heavy, medium and field regiments plus anti-aircraft guns
agst	against
AFV	Armoured Fighting Vehicle
Amn	Ammunition
AOP	Air/Armoured Observation Post
AP	Armour-piercing or anti-personnel
APCBC	Armour-piercing ballistic capped
APC	Armoured Personnel Carrier
APDS	Armour Piercing Discarding Sabot, 6- and 17-pounder anti-tank ammunition
APIS	Air Photo Interpretation Section
Arty	Artillery
Assy Area	Assembly Area, where troops regroup (e.g. tanks, infantry and gunners) for an operation prior to moving to the FUP
Aslt	Assault
A/Tk	Anti-tank (+ variants)
ARV	Armoured Recovery Vehicle
AVRE	Armoured Vehicle Royal Engineer
Axis	The central direction of advance

BC	Battery Commander
B Echelon	The entry point of combat supplies into a unit's logistic infrastructure
BCR	Battle Casualty Replacement
Bde	Brigade (British) equivalent to a US or German regiment
BK	Battery Captain
BGS	Brigadier General Staff
BM	Brigade Major
BMA	Beach maintenance Area (Logistic dumps)
Bn	Battalion
Br	Bridge
BSM	Battery Sergeant Major
Bty	Battery
CAS	Close air support
Cas	Casualties
Casevac	Casualty evacuation
CCP	Casualty Clearing Post
CCS	Casualty Clearing Station
Cdn	Canadian
C in C	Commander in Chief
Civ	Civilian
CL	Centre Line or axis of advance
CO	Commanding Officer*
Co-ax	Coaxial machine gun mounted in tanks alongside the main armament
Conc	Concentrate also Concentration Area – where units assigned to an operation gather before moving forward to the assy area
Comd	Command
Coord	Coordinate
Coy	Company
CP	Command Post
Cpl	Corporal
CPO	Command Post Officer – artillery
CRA	Commander Royal Artillery (divisional)
CCRA	Commander Corps Royal Artillery
CS	Close Support
CSM	Company Sergeant Major
D Day	The day on which an operation begins
Def	Defence
Dem	Demolition
Dets	Detachments

*Today the distinction between CO and OC is well-defined, with the former commanding a unit such as a battalion and the latter a sub-unit such as a company, but in many British Second World War memoirs and documents they are transposed or used inconsistently.

DF	Defensive fire, pre-planned artillery, mortar or medium machine gun target areas on likely enemy approaches, etc
DGNS	Dragons
Div	Division
DP	Distribution Point, logistic
DR	Dispatch rider, also 'Don R' or 'Don Romeo'
DZ	Drop Zone
Ech	Echelon, normally referring to fighting echelon and logistic echelons of a unit
Fd	Field battery, company or squadron RE or RA
Fmn	Formation, i.e. brigade, division, etc.
En	Enemy
Enfilade	Fire from a flank
Eqpt	Equipment
Fd Regt	Field Regiment RA
FMC	Field Maintenance Centre
Fmn	Formation, normally referring to brigades or divisions
FOB	Forward officer Bombardment – naval gunfire
FOO	Forward Observation Officer – artillery
Est	Establish
FUP	Forming Up Place
Fwd	Forward
Gd	Guard typically advanced, flank or rearguard
GOC	General Officer Commanding
GPO	Gun Position Officer
Gr	Grenadier
H Hour	The hour (time) at which an operation starts
HE	High Explosive
Hy	Heavy
Inf	Infantry
Int	Intelligence
I Sec	Intelligence Section
i-tank	Infantry tank
JNCO	Junior Non-Commissioned Officer
KIA	Killed in action
KO	Knocked out
LAA	Light Anti-Aircraft
LAD	Light Aid Detachment
LCA	Landing Craft Assault
LCpl	Lance Corporal
LCT	Landing Craft Tank
Ldr	Leader
Line	Field telephone wire
LSI	Landing Ship Infantry

LZ	Landing Zone
MG	Machine gun
MIA	Missing in Action
MMG	Medium machine gun
MSR	Main Supply Route
O Group	Orders group, assembly of subordinate commanders for the issue of orders
OC	Officer Commanding
OP	Observation post
OP O	Operation Order
P Hour	Time of a parachute drop
PIAT	Projector Infantry Anti-Tank
Pl	Platoon
Posn	Position
Pt	Point
Pte	Private (rank)
PW	Prisoner of War
Pz	Panzer
QM	Quartermaster
RAP	Regimental Aid Post
Rd	Road
Regt *or* Rgt	Regiment
REME	Royal Electrical and Mechanical Engineers
RHA	Royal Horse Artillery
RHU	Reinforcement Holding Unit
RMO	Regimental Medical Officer
RPM	Rounds per minute
RSM	Regimental Sergeant Major
RSO	Regimental Signals Officer
RV	Rendezvous point
SA	Small Arms
SB	Stretcher bearer
Sec *or* Sect	Section
Shelldrake	Appointment title for artillery
Shellrep	Report of enemy shelling. Used to locate enemy batteries
Sgt	Sergeant
Sitrep	Situation Report
Sjt	Serjeant, the old form of sergeant still used by some regiments
SL	Start Line
Smk	Smoke
SNCO	Senior Non-Commissioned Officer
SP	Self-Propelled
Sp	Support

Stand-to	'Stand to arms', fully man defensive positions, routine at dawn and dusk.
Sub-unit	Company, squadron or battery-sized unit
Sunray	Commander of any-sized unit of formation
Sups	Supplies
Sqn	Squadron
Svy	Survey
Tac	Tactical, usually used in context of a tactical headquarters
Tac/R	Tactical reconnaissance by aircraft
Target	All guns within range when a '*xxxxx* target' was called: Mike (regiment), Uncle (division), Victor (corps), Yoke (army)
TCV	Troop-Carrying Vehicle
Tgt	Target
Tk	Tank
Tp	Troop
Tpt	Transport
Unit	An infantry battalion or a British artillery or armoured regiment
VCP	Visual Control Party, RAF and gunner forward air controllers
Wef	With effect
WIA	Wounded in action
X Rds	Crossroads
Yd/yds	Yards
Yeo	Yeomanry

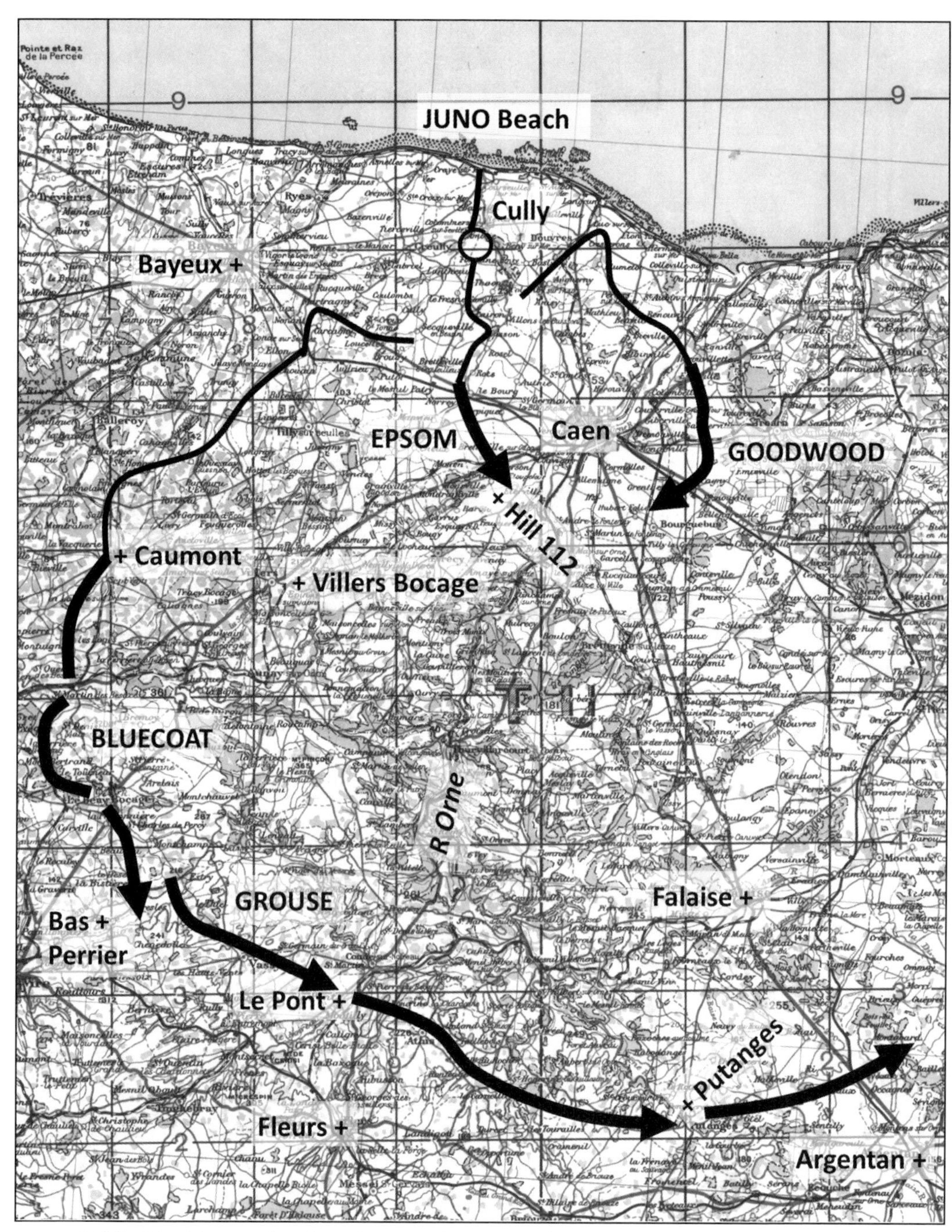

8 Rifle Brigade's operations from their arrival in Normandy to the closing of the Falaise Pocket.

Chapter One

In the Beginning

Britain had been a leading light in the development of the tank during the Great War, but following the Armistice there was only modest armoured development for almost a decade. This was against a background where trench warfare and all it involved was regarded as aberrant, and the regular army settled back into 'real soldiering' out in the Empire. Meanwhile, abandoned tanks rusted away on Bovingdon Heath. This, however, did not stop theorists, including Liddle-Hart, Fuller, and Hobart, from working on a variety of armoured issues. Meanwhile, in Europe, vast sums were spent on elaborate border fortifications, the antithesis of armoured warfare.

However, with the example of the utility of armoured cars and aircraft working in tandem in the Middle East and a poorly handled army exercise during 1925, the British Army eventually established a brigade-sized Experimental Mechanised Force (EMF). Based at Tidworth on Salisbury Plain in the spring of 1927, the EMF trained in the practicalities of armoured warfare alongside the development of doctrine, all of which was tested during a series of often two-sided exercises.

A long list of equipment and vehicles to make the EMF an effective armoured formation resulted, but with the rolling no-war-for-ten-years policy and the world sinking into the Great Depression, the shopping list was unaffordable. Among the casualties was the clearly articulated case for a vehicle that would enable the infantry to keep up with the tanks across country. The concept of an all-arms armoured formation was abandoned, although much of the doctrine produced mirrored that subsequently developed during the Second World War. Instead, a tank brigade was formed in 1933. Over the following years, this brigade's doctrine fluctuated between infantry support and independent action as commanders changed and theorists continued to divide over the employment of armour.

In 1935, with a resurgent Germany under Adolf Hitler shrugging off the restrictions of the Versailles Treaty, the Army's general staff recognised the probability of renewed war. They increased the pace of general mechanisation and started work on a mobile mechanised division. Battling opposition to rearmament, it was not until 1937 and growing German adventurism that Mr Churchill's warnings of coming war could no longer be ignored. A formal establishment for a new mechanised infantry role was issued in 1938 and the following year the Mechanised Division became the 1st Armoured Division.

The EMF proved that without a carrier of some kind that would enable the infantry to keep up with tanks, the all-arms concept did not work properly.

The Rifle Brigade's historian explained the requirement that brought the Rifle Brigade (RB) and King's Royal Rifle Corps (KRRC) into a specialised role:

> It was with this background of events that the plans for forming an armoured division were implemented. An early decision was taken that tanks would require specially trained infantry to work with them in close support. These infantry would have to be fully mobile so that they could keep up with the armoured advance. Their main role would be to restore the momentum of the attack when the armour was temporarily held up by the enemy in built-up areas or in rough or mountainous country where the tanks could not operate by themselves. Many subsidiary tasks were to emerge later.[1]

Vickers machine guns and 4.2-inch medium mortars had already been allocated to four infantry regiments. Similarly, with the anticipated expansion of armour, the primary providers of motor battalions to armoured brigades were the KRRC and RB. There were exceptions to this rule overseas, where line battalions took on the 'motor' role along with the Grenadiers' 1st Battalion in the Guards Armoured Division. The motor battalion specialism initially involved only the regiments' regular army battalions, but in the aftermath of the salutary 1940 campaign, most of the RB's and KRRC's Territorial Army battalions also became motor battalions during the wholesale raising of armoured formations.

The 2nd Battalion, The London Rifle Brigade

The 8th Battalion, Rifle Brigade that fought with the 11th Armoured Division in the North-West European Campaign had its antecedents among the many London rifle companies at the birth of the Volunteer Movement of 1859–60. In the Cardwell Reforms, the volunteer battalions were grouped into the London Regiment, and in the post-Anglo-Boer War 1908 Haldane Reforms, the battalion was incorporated into the Territorial Army as the 5th (City of London) Battalion, The London Regiment. When the Territorial Army was reformed after the First World War, the battalion's title was 5th City Battalion, The London Regiment (London Rifle Brigade).

The 8th Battalion that served during the Second World War is, however, not to be confused with the 8th (Service) Battalion of the Rifle Brigade that was raised as part of Kitchener's New Army in 1914 and disbanded in August 1918. The battalion we will be following in these pages was formed in east London in 1939 as part of the doubling of the Territorial Army, as the 2nd Battalion, The London Rifle Brigade.

The 2nd London Rifles (as they were colloquially known) were mobilised on 1 September 1939 when Hitler invaded Poland. As part of the 5 London Brigade of the 2nd London Division, the battalion manned roadblocks in the anti-invasion role before returning to billets in east London. Remaining in the city close to the distractions of home and hearth was unsatisfactory, and the division joined others across the army in moving away from such diversions. This meant a move to Eastern Command before a move further from London to Western Command.

The London Rifle Brigade's cap badge. This was gradually replaced by that of The Rifle Brigade as riflemen arrived from the Motor Training Battalion or the badges' slider broke.

When the division was placed on a lower establishment, in January 1941, the 1st and 2nd battalions of the London Rifles respectively became the 7th and 8th battalions of The Rifle Brigade, the full mouthful of the title of the battalion we are following being, 8th Battalion, Rifle Brigade (Prince Consort's Own) (London Rifle Brigade). As far as companies were concerned, with the 7th Battalion retaining the designations A to D, the 8th Battalion's companies kept E to H.

11th Armoured Division

With the all-too-stark lessons of the 1940 Campaign having been digested for home defence and ultimately to fight the Germans on their own terms, Britain invested heavily in armoured formations. By December 1940, plans were made for the raising of the 11th Armoured Division, but it was not until March of the following year that the divisional headquarters was formed in Yorkshire, with Major General Percy Hobart charged with training the division. He had recently returned from the Middle East, where he had raised the 7th Armoured Division, which was only the latest of a long line of achievements since he transferred to the Tank Corps in 1923. There was no better officer to start the process of taking the units identified to make up the 11th Armoured Division's order of battle.

As a battalion of a regiment earmarked for the motor role, 8 RB left what had become the 47th (London) Infantry Division and travelled north to join the 29 Armoured Brigade, which at the time was one of two such brigades, and the 11th Support Group that made up the division. The divisional historian describes the inspiration for this ORBAT and some of its disadvantages:

> For the organisation of the new British armoured formations was designed to afford the maximum number of tanks at the expense of any other arms. Its two armoured brigades, with three armoured regiments in each, contained some 350 of these machines. Yet in the whole division there were only 24 guns capable of supporting all these units; and the infantry too was very short, three battalions in all. The choice of such an organisation was inspired by the dominant and decisive part played by tanks in the German victories and the consequent belief that an armoured division had but two clear-cut roles; firstly, to seek out the enemy's armour and destroy it – for which purpose infantry would be, superfluous – and, secondly, to exploit a breakthrough – in which case possible counter-attacks would only come from the enemy's lines of communication troops, and to repel these all that was required was a small number of well-armed and highly mobile infantry who should be able to travel close behind or with the tanks, and take over successively the pieces of ground which their advance had gained.[2]

Major General Hobart proved to be a hard taskmaster as he set about welding the majority of Territorial Army units and war-raised regiments, leavened by a handful of Regular Army units, into a division that was arguably to become the most effective armoured formation in the North-West European Campaign.

Major General Percy Hobart, the division's first general officer commanding. (Inset) The sign of the 11th Armoured Division. The bull was part of General Hobart's family crest.

The 11th Armoured Division training with Valentine tanks on the North York Moors in late 1941.

Over the cold east coast winter of 1941–42, the 8th Battalion joined the rest of 29 Armoured Brigade concentrated in the Whitby area before heading south to Surrey and Sussex for three months of exercises, including TIGER, which was run by the GOC of Southern Command, one Lieutenant General Montgomery.

Experimentation on the North York Moors, the 1942 exercises and reports of armoured warfare from the desert soon saw 30 Armoured Brigade leaving the division.[3] This brigade, as others, moved to the short-lived 'Mixed Infantry Divisions' but in this process the 11th Armoured gained 159 Brigade as its lorried infantry, from the 53rd Welsh Division. The divisional historian explained the rationale behind this change of ORBAT:

By the spring of next year (1942), the results of the fighting in the Middle East had demonstrated that an armoured division should be and had to be employed on tasks other than battles to the death against the enemy's armour and long-range exploitations of breakthroughs. The Germans had reduced the armoured component of their panzer divisions and increased the infantry; and a similar reorganisation was now decreed for British formations. One of the two armoured brigades was to be given up and a lorried infantry brigade substituted. The acquisition of this enabled the armoured division often to regain the momentum of its advance when slowed down or actually halted by enemy defensive positions; to establish bridgeheads over rivers, to fight effectively in close country; even on occasions to mount a pitched battle of its own. The overall effect of the change, in fact, would be to increase the number of armoured divisions and make those divisions more versatile.

Now based in equally cold East Anglia, in early 1943 the division was warned for overseas service, which most correctly assumed would be the deserts of North Africa. With the advance party having sailed and much of the rest of the division embarked on shipping, the deployment was cancelled, as the terrain in Tunisia was more suitable for an infantry division.

With no prospect of operations in 1943 for the division and its motor battalion, it was a return to the round of exercises ranging from company to corps. Still in East Anglia, much of this period was spent on Stanford Training Area, with the dominating central feature of Frog Hill becoming all too familiar to the riflemen, as it has continued to be to generations of their successors.

The Motor Battalion

While the organisation of British armoured divisions changed throughout the early years of the war, the envisaged role of the armoured division altered little but was heavily influenced by the 1940 Campaign and desert warfare. The armoured division's role was essentially to fight a mobile battle and lead the pursuit. In contrast, tank brigades equipped with the heavier and slower-moving Churchill tank, which was originally conceived as an i-tank, were organised to supported infantry formations during break-in battles.

As was the case with most areas of the British Army's tactical doctrine during the years after 1940, that for the motor battalion underwent numerous changes before the definitive military training pamphlet, MTP No. 41, was published in June 1943. Its opening paragraphs cover the 'Principles of Organization':

1. The motor battalion forms an integral part of the armoured brigade and provides the brigade with the immediate and close support of troops capable of carrying out those mounted infantry tasks for which the organization, equipment, and armament of armoured regiments are unsuited.
2. The personnel of motor battalions are therefore primarily infantry and fight on their feet. Their unarmoured vehicles are a means of transport only and will never be used as armoured fighting vehicles. They must, however, be used boldly and, in order to give the armoured regiments as close support as possible, particularly in the deliberate attack, personnel should be carried in their vehicles as far forward as is consistent with the safety of the crews.
3. Armoured carriers and scout cars, though 'A' type vehicles, are not true armoured fighting vehicles, and it will be the exception for them to be used as such, although favourable moments may occur, particularly in confused fighting, for mounted action.

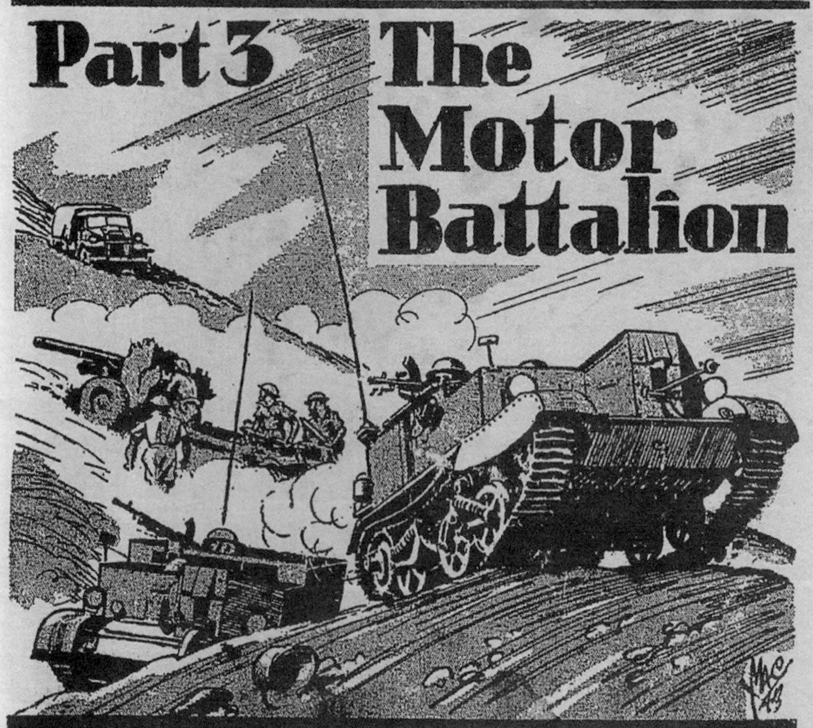

The front cover of MTP No. 41.

A motor battalion still equipped with 15-cwt trucks for mobility on parade for inspection.

4. The organization of the motor battalion is designed to permit the maximum mobility and flexibility; each company contains its own reconnaissance and administrative elements and is thus a self-contained unit.[4]

As will be seen later in this book, the reality of fighting the Germans who had been ordered to rope off the Allied lodgement in Normandy, caused doctrine that emphasised exploitation and pursuit to be reconsidered in the light of experience and the bocage country.

Equipment

As with manpower, shortfalls in weapons, equipment and vehicles were made up as the invasion approached. The war diary noted as late as 24 May: 'Machine gun carriers arrived to replace carrier universal as M.G. carrier in E Coy.' The main late arrival on 24 April were half-tracks, which replaced the White scout cars in battalion and company headquarters.

Final Training and Preparations

During December 1943 veteran officers who had returned from the Mediterranean with General Montgomery started to arrive in the division. Chief among these changes was Major General Pip Roberts, who replaced General Burrows in command of the division. Closer to 8 RB was their new brigade commander, Brigadier Harvey, who was described as 'a thrusting cavalryman'.

On 3 January 1944, the 11th Armoured Division returned to North Yorkshire, with 8 RB occupying a camp near the coastal village of Hunmanby. Here the division was to conduct a series of final exercises but at the end of the month the motor companies travelled by train back south to the blitzed East London Docklands of Limehouse. During five days of urban warfare training (Exercise CHINATOWN), as was often the case, officers and men were wounded and injured, some seriously, during live firing exercises that were deliberately as realistic as possible.

Meanwhile, even though by now 8 RB knew that they would not be taking part in the invasion's initial assault, vehicle drivers and commanders were sent to practise embarking and disembarking on landing craft tank (LCT).

Major General 'Pip' Roberts, commander of 11th Armoured Division.

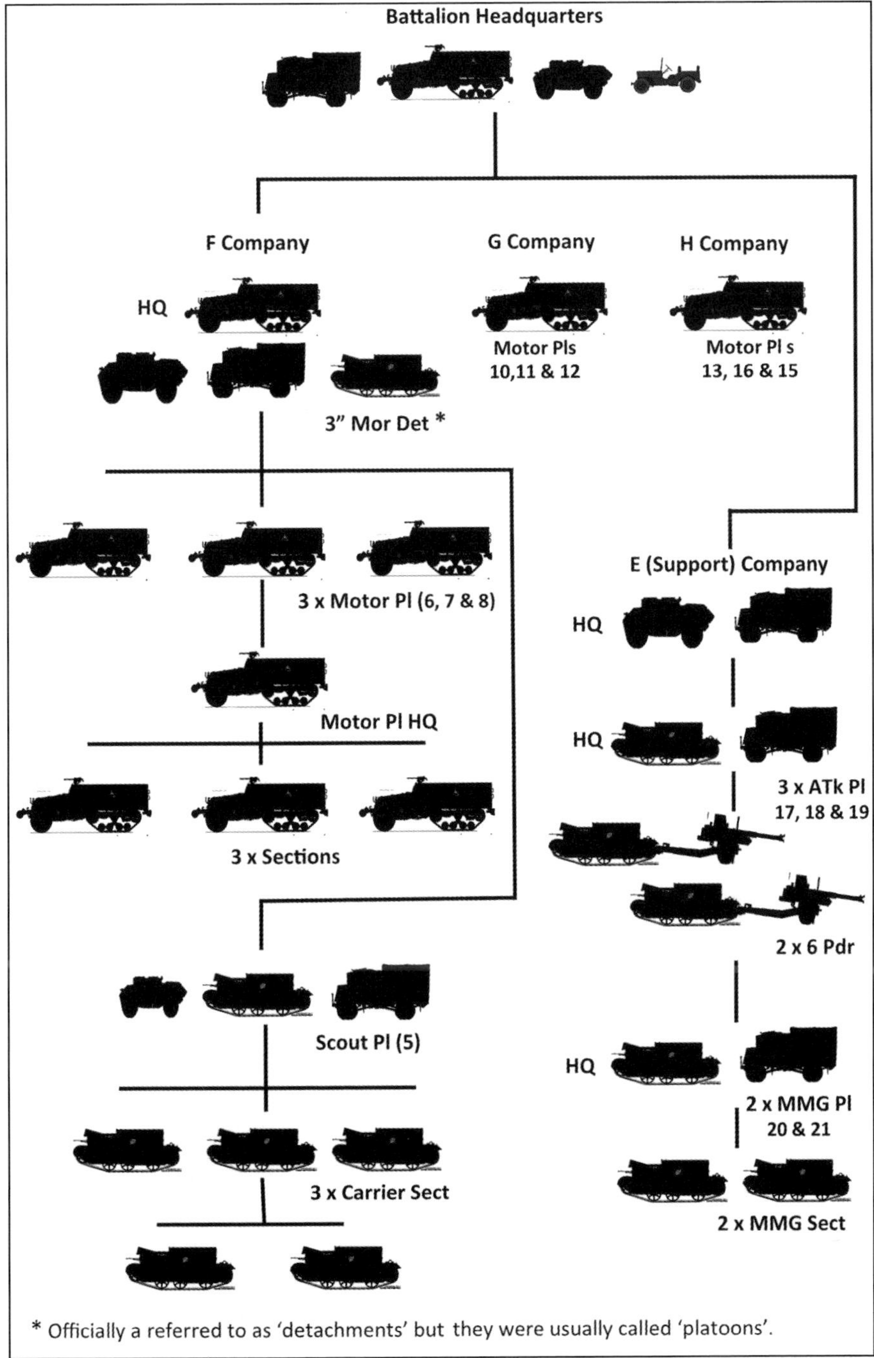

The organisation and main equipment of 8th Rifle Brigade.

The Half-Tracks

Before the war the armoured brigade of a division had an infantry battalion in its order of battle, or as it became known, a motor battalion. The infantry were transported in trucks of various types with invariably a limited cross-country performance and, therefore, having no great ability to keep up with tanks.[5] By 1943 the establishment tables showed motor battalions as having 'Trucks, 15-cwt – 4 × 4, personnel' with a better cross-country performance but, of course, no armoured protection. These vehicles were, however, at the time already being replaced by the American-manufactured Scout Car M3A1, known in British service as the White scout car, but like the 15-cwt truck, it was too small for a full rifle section. Consequently, the decision was made in preparation for the invasion that they and the 15-cwt truck would be replaced in motor battalions, as well as in certain tactical and sub-unit headquarters, along with miscellaneous units that were mainly armoured. At short notice, a variety of US 15-cwt half-tracks were procured under lend-lease but those received excluded the M3 half-track that was equipping US forces. Of the 2,202 (not including the war maintenance reserve) needed for 21st Army Group's units, a considerable proportion were a mix of International Harvester Corporation (IHC) M5s, and from a number of other companies that turned to war production. The M5, M6 and M9 were of a similar design.

Thanks, however, to shortages of these three types, the M14 anti-aircraft version with its Maxson M33, mounting twin M2 .50cal machine guns, was converted by removing the weapons mount, plating over the hole in the deck and cutting a

The IHC M5 half-track.

door at the back of the vehicle's rear compartment. Seating for eight to ten was normal in half-tracks issued to the British infantry. Most units including 8 RB had a mix of types of half-track at the opening of the campaign. However, they all provided protection from small arms, mortar and shellfire but the 1/4 to 5/16-inch steel provided little protection from direct hits by HE or AP fire.

The US-produced half-tracks could outperform the German equivalent, the Hanomag Sd.Kfz.231, in almost all aspects, including speed and turning circle, with the exception of a marginal carrying capacity. The US M-series of half-tracks had a shorter track base and power steering, which together greatly improved performance.

The new commander of VIII Corps, Lieutenant General Sir Richard O'Connor, presided over the last corps exercise, which proved to be a challenge for men and equipment out on the North York Moors:

A long and final exercise with troops called 'Eagle' followed towards the end of February, which certainly proved a test of endurance and of fitness for battle. Starting off with dispositions as in a bridgehead, 8 Corps was launched in the usual break-out role in the most severe weather. Several of the opposing troops dug-in and awaiting the attack, are said to have died from exposure in the Arctic conditions prevailing. Every arm and service was employed in as realistic a representation of winter campaigning as the most enthusiastic writer of military exercises could desire, and 8 Corps came through with flying colours.[6]

Back in camps, during the first three months of the year the division received numerous visitors and 8 RB regularly found itself on parade. These included a visit by General Montgomery, who looked at rather than inspected the riflemen before delivering one of his informal chats. The visits culminated with the King, Queen and Princess Elizabeth inspecting 29 Armoured Brigade on 22 March.[7]

On 30 March the battalion departed for a 'new location', leaving F Company to follow once they had completed their range package. With security steadily mounting, the battalion's motor companies along with the rest of the division began a two-day drive south to Aldershot and the armoured concentration area for build-up formations. From here tanks, vehicles, including 8 RB's half-tracks, and the infantry of 159 Brigade would, when the time came, all go to separate embarkation ports and reassemble in Normandy.

Once at Aldershot the number of troops in the area meant that access to training grounds was limited, and anyway the all-consuming task of water-proofing the battalion's vehicles for a deep wade from landing craft to the beach kept everyone occupied. This was defined as being 6ft and had to be done in stages, and the final waterproofing, which would result in engines overheating, had to be carried out at the point of embarkation or aboard shipping. The results of the driver's labours were tested at the Royal Engineers' hard at Hawley Lake.

The King's inspection of 8 RB at Bridlington.

Manning the Battalion

The origins of the 8th Battalion as a London Territorial Army unit had in the four and a half years' service in the UK before being deployed in Normandy been steadily watered down by the posting in of conscripts, who thanks to the infantry manning system could come from across the country. However, probably because of the motor battalion specialisation, 8 RB was largely spared General Montgomery's programme of exchanging large numbers of officers and senior NCOs with veteran units that had returned from the Mediterranean. The commanding officer, Lieutenant Colonel Treneer-Michell, a territorial officer, remained in place, as did the company commanders, but by the end of January 1944 it was noted in the war diary that: 'There are now in the Bn about 30 men who have the Africa Star for service with the 8th Army.' These were NCOs from the 1st Battalion who had served with the 7th Armoured Division. Also from January onwards, the battalion was brought up to its full war establishment with junior officers and men being posted in from the regiment's 2nd Motor Training Battalion (MTB), based at Ranby in Nottinghamshire.

8th Rifle Brigade's Numbering of Platoons

The numbering of platoons in motor battalions did not follow that introduced in 1943 for the standard infantry battalion:

Battalion Headquarters

Headquarter Company	Company HQ, 1 Signal Pl, 2 Administrative Pl, MT, REME LAD, RAP, QM's
E (Support) Company	17, 18 and 19 anti-tank pls, 20 and 21 MMG pls
F (Motor) Company	Company HQ and 3-in mortar det 5 Scout Pl, 6, 7 and 8 motor pls
G (Motor) Company	Company HQ and 3-in mortar det 9 Scout Pl, 10, 11 and 12 motor pls
H (Motor) Company	Company HQ and 3-in mortar det 13 Scout Pl, 14, 15 and 16 motor pls

The battalion's establishment of 38 officers and 816 other ranks was, despite having only three bayonet companies, the same as a standard infantry battalion. The necessity for additional drivers, commanders, their own medium machine gun platoons and an enhanced admin and repair staff for an armoured unit accounts for this.

One significant departure was recorded in the war diary during February: 'RSM A.H. Edwards who had been RSM of the Bn since the beginning of the war was downgraded and posted to 2 MTB.' Officially the age limit of 36 imposed by General Montgomery was for commanding officers but in practice it extended to other key personnel. In this case medical downgrading precluded service outside the UK. Another departure recorded on 7 May was '1st Reinforcements (five officers and sixty-five other ranks) left for a Reinforcement Holding Unit [RHU]'. These men, alongside others from the rest of the division, were held at VIII Corps' RHU and once in Normandy fed forward as individual reinforcements as required. These men, unlike later replacements, had served in the battalion, knew its ways and were well trained. Another 10 per cent of the battalion were the 'left out of Battle' (LOB) and would deploy to Normandy with the echelon, as a source of immediate replacements. Most officers and men were unhappy at being cast in these roles but after less than two weeks the LOB men were back with their companies following the division's first battle.

Deployment

The 11th Armoured Division was 'phased in' early in the build-up, originally being planned to land in Normandy between D+4 and D+6 but from the outset the weather in the Channel had delayed landings and the loss of equipment such as Rhino ferries had caused mounting delays. Consequently, even though the division was called forward by the Build-up Control Organisation (BUCO) on 8 June, there were delays for the main body of 8 RB in the marshalling area and

The booklet issued to British troops while in their embarkation camps, along with French francs.

aboard ships. On D+3 they were on the road from the Aldershot concentration area to the marshalling area in east London at 0530 hours. Rifleman Patience from the East End recalled the journey through familiar streets to a dismal transit camp at Purfleet and the docks:

> We were on our way – but little did we all know that we were to go through parts of London where we lived. One of my mates, who joined up with me, saw his mum in Leytonstone. We all looked a sorry sight by now because

the fumes of the engines were coming into our vehicles, making us look as if we were crying. We came down the A12 through Wanstead, Gants Hill and Whalebone Lane, and then there I was, passing the fields that, as a kid, I used to play on. I could see the ponds that we used to fish for newts and tadpoles.

Next came Mawney Road roundabout, only a few yards from home ... To my surprise, I saw my sister Kath standing by the post box only a few yards away. I am sorry to say that she didn't see me – maybe it was for the best though. We then passed North Street, Pettits Lane, Gallows Corner and all of a sudden it was gone.

I remember thinking to myself, 'Will I ever see this again? Maybe I will or maybe I won't.' One thing we all knew was that a lot of us would not be coming home.[8]

During 9 June the battalion was at work in the docks at Tilbury carrying out the final stages of waterproofing that could only be completed at the dock for mechanical reasons. Rifleman Patience continued:

We were really annoyed that the dockers were on strike and refused to load our transport. Our own engineers and the ship's cranes had to do the job. The names we called those dockers ... well they are unrepeatable. As we went to board the ships the people of Tilbury lined the streets to bid us farewell. They did all right because we threw all our spare change to the children. The ship, an old Yankee ship called *Samsit*, was really dirty. We didn't know what it had been used for before, but the holds stank of all sorts. We had hammocks to sleep in and it was also the start of many months of not removing our clothes.

On 10 June the battalion finally left the transit camp for embarkation. Aboard two ships, officially known as MT 88 and MT 107, the battalion left the docks at 1530 hours for a two-hour passage down the Thames. Once out in the estuary, they joined the assembling convoy anchoring off Southend.[9]

With the build-up by now running approximately a day and a half behind schedule, the Royal Navy and BUCO were constantly working on the mammoth task of reworking their tables and allocating shipping to units. After a day waiting in the confines of the ships, with platoons taking it in turns to use the little open deck space for PT, the first of the convoys bearing the 11th Armoured Division slipped anchor on 12 June at 0600 hours. Rifleman Kelley of F Company wrote of the voyage to war:

Monday morning awakened us with a rude shock. We were on our way! As a driver would say, 'We went straight down the Thames, turned right, and turned right again at Dover.' As we passed through the Straits, a smoke screen was laid to give us cover from Jerry guns on the coast of France. Towards nightfall, after doing our umpteenth Emergency Stations act, we neared the Isle of Wight, and there was nothing else to do but to retire to the

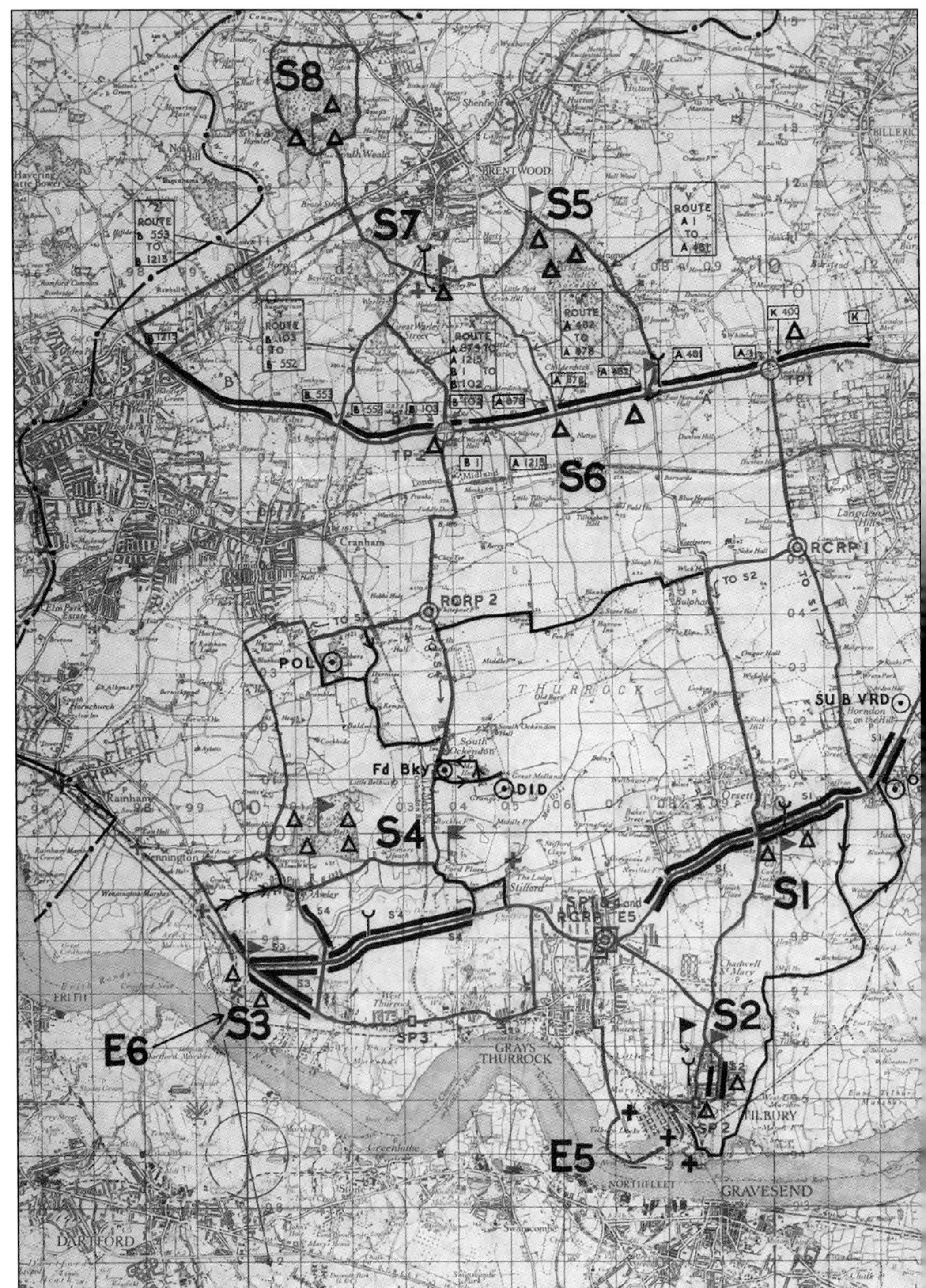

Area S was where 8 RB embarked (E5) for Normandy. The mounting of the operation from the UK was a hugely complicated part of Operation OVERLORD.

Infantry disembarking on JUNO Beach in the days after the invasion.

Blackhole for some sleep. For who knew when he would be able to sleep again? Before sleep arrived, imaginations ran loose on the morrow's events. Jerry planes bombing the convoy? Jerry shells? U-boats? Who could tell? Once in the night the deep roar of depth charges filled the still night air, which left men with cold sweat on their faces. All those awake lay tensely in their huddled positions. Not even the captain of the boat could help them now if a torpedo struck. However, the rest of the night was uneventful.

Old and dirty the transports may have been, but as American ships they had one big advantage for soldiers whose rations had steadily been reduced over four years of war; food was according to the war diary, 'unlimited'. Captain Straker recorded the arrival of H Company off the coast of Normandy on 13 June:

We had been on the two ships, which had carried the Battalion, for nearly a week, and so everyone was up early to get a first glimpse of the country we had come to, the Normandy Beachhead. The sea was black with ships and we were all surprised that we were left well alone by the Luftwaffe to disembark at our ease. We were anchored about 3 miles off the shore. At last, at about midday, the first tank landing craft moored alongside the ship. Vehicles were picked out of the hold by the ship's derricks and lowered into the LCT. The crews had to scramble down a rope ladder. As each LCT was loaded it set off for the shore.

The No. 4 Lee-Enfield Rifle

The rifle was carried in battle by six of the eight dismounts in infantry sections of a motor battalion. The weapon used by the riflemen of 8 RB had its origin in the aftermath of the Boer War that had exposed sundry deficiencies in the British Army's weapons and equipment. In 1904 the .303 No. 3 Short Magazine Lee-Enfield Rifle (SMLE) was issued across the army and represented a significant

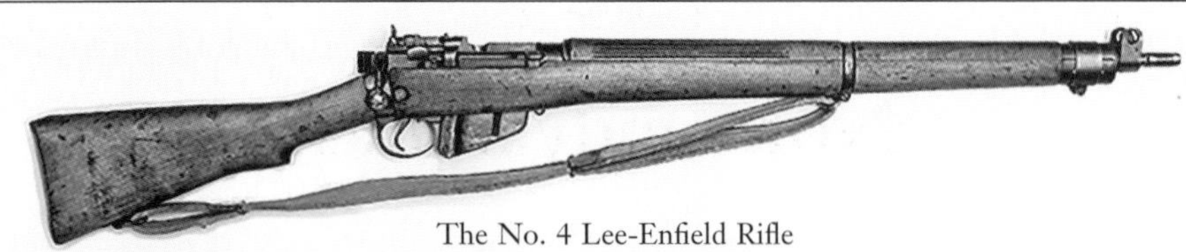

The No. 4 Lee-Enfield Rifle

improvement of its predecessor, incorporating features from other modern weapons. With the salutary lesson of Boer marksmanship, all arms of the army trained hard in its use and, consequently, the SMLE was a significant factor in the success of the BEF in 1914. During the war the design was simplified largely for manufacturing reasons.

Post First World War, the No. 3 Rifle remained in service with further refinements until replaced, starting in 1941, by the new but similar No. 4 Lee-Enfield Rifle. This weapon was further simplified for wartime mass-production and issued from 1942 onwards. By 1944 most theatres had received the No. 4 Mk I.

The long bayonet issued with the SMLE, necessary for bayonet fighting with the longer barrelled rifles of most European armies prior to the First World War, was replaced with a much shorter spike bayonet.

The Riflemen of 8 RB were taught to shoot out to 300 yards as individuals and to 600 yards as a section.

With the bolt open, a five-round stripper clip was fitted to the charger guide for speedy reloading of the ten-round magazine.

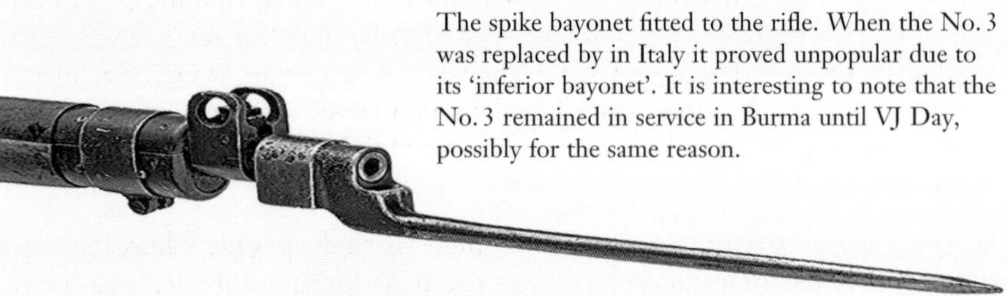

The spike bayonet fitted to the rifle. When the No. 3 was replaced by in Italy it proved unpopular due to its 'inferior bayonet'. It is interesting to note that the No. 3 remained in service in Burma until VJ Day, possibly for the same reason.

Chapter Two

Normandy at Last

The division's main body crossed to Normandy between 13 and 17 June, with 8 RB being the first unit of 29 Armoured Brigade to land. Brigadier Harvey's logic was that, being on the ground ahead of the tanks, the battalion would be able to defend the brigade's concentration area if it was under threat.

At 1200 hours on 13 June LCTs began shuttling 8 Rifle Brigade to MIKE and NAN Sectors of JUNO Beach, astride the small port of Courseulles. The Royal Navy beach master's officers and ratings called the craft in to the next available stretch of beach. Captain Noel Bell described the journey into the beach:

> We looked ahead of us. The coastline became more distinct as we approached and we perceived there were sand-dunes at the top of the beach in front of the village. Behind the latter the ground rose steadily to form a low ridge which blotted from view the country beyond. Estimation of distance over water is always deceptive, and we found we had farther to go than we had imagined; but finally, anticipation gave way to reality. The bottom of the craft grated against the shingle – it shuddered, then came to a standstill. The ramp went down. Before us lay Courseulles-sur-Mer.

The first craft bearing 8 RB's vehicles touched down at 1505 hours. Captain Bell noted that 'We were there.'

> The trucks [half-tracks] drove off into the water, but the pilot had found a place that gave us, as near as possible, a dry landing, and the sea did not encroach inside the vehicles to any extent. A naval officer was saying, 'Keep her revving; don't get stuck there.' We kept her revving. The engine hic-coughed, and we felt the tracks make the gradient of the beach. We emerged. Water dripped off the suspensions. The sand, soft as it was, felt firm and good after our seventy-two hours afloat. A military policeman was pointing where to go. We turned left and down along the beach, running parallel to the sea. The barbed wire hung in torn shreds, and the sand-dunes were pocked with shell-holes and slit trenches. On a corner where we turned right, down a track, there was a little wooden cross, made out of a 'compo' box, inscribed 'A Canadian soldier lies here'. That was all. No name, no regiment, no date.

Having laboured so hard to prepare the vehicles for a deep wade ashore, landing in just a couple of feet of water, most drivers had mixed feelings of relief and curiosity to see if their work would have been successful. In F Company's case 'some

Soldiers of the Beach Group working on an exit and looking on as tanks and infantry land on JUNO Beach.

vehicles didn't even get their tyres wet'. Once across the beach the first task was to remove sufficient of the so carefully applied waterproofing to allow the battalion's vehicles to clear the beach area without damage to engines or running gear.

Units and headquarters across the division had sent advance parties on ahead that had landed the previous day to sign routes from the beaches and establish harbours in the divisional concentration area. All the vehicle crews knew was that they were to follow the signs and directions of the military police through the hive of activity that was Beach Maintenance Area STARS. The route the battalion was supposed to follow led south through Creully to the village of Cully. For most of the riflemen the journey was the first experience of the sights and sounds of war. Rifleman Patience, with the rattle of machine gun fire audible in the distance, remembered:

> the terrible smell – the stench of cordite burning buildings but most of all the stench of dead animals, hundreds of them lying dead on their backs. It was a hot day and the smell was terrible – never have I seen such terrible slaughter of animals. I felt sick.

The first elements of the battalion's main body reached the harbour area in the orchards north of the village of Cully at 1800 hours but, as noted in the war diary, stragglers were still coming in the following day. Some of these were the last vehicles of F Company to be landed, having spent a very uncomfortable night on an LCT. Most were, however, the result of breakdowns, mis-direction and the missing of turnings, which more often than not resulted in long journeys into the unknown on slow-moving, one-way traffic circuits. For those who successfully followed the signs, it seemed all so simple but the historian of fellow members of 29 Armoured Brigade, the 23rd Hussars, explained the pitfalls:

> They set off into the 'unknown' though everyone knew very well what the procedure should be … The explorers had found their way. They had left signs so that those that came after might know where to go. So many had explored. Arrows pointed every way and in the end it was by devious routes that the vehicles eventually found their way to the Regimental concentration area … Some followed the right arrows. Others of a more independent turn of mind chose their own routes and hoped for the best.[1]

Meanwhile, most of G Company had reached Cully:

> Our little convoy turned into a farm, proceeded through the courtyard and into the orchard beyond. We drove right round the orchard, keeping to the sides … We halted, parked the trucks in some form of order along the hedge, dug in again, and put up camouflage. The carriers arrived later, and by the time they were in it was nearly dark. We made our beds close by our slit trenches, and put our rifles, steel helmets and equipment beside them; we were new to the game and wished to take no chances, although the nearest Boche was reported over 3 miles away.

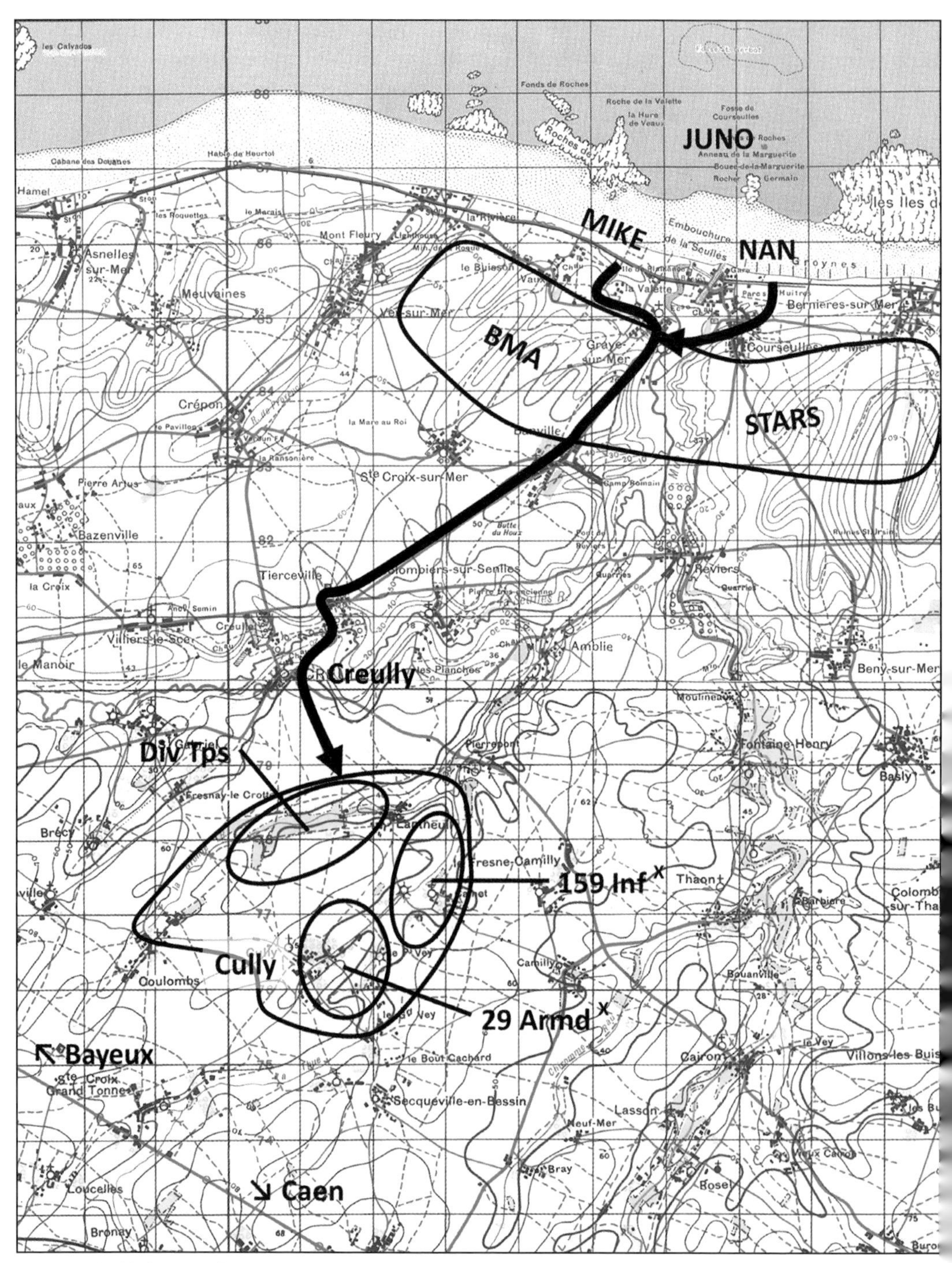

The likely route from MIKE and NAN beaches to the divisional concentration area.

We took off our jackets and boots and slid in between our blankets. The evening was still, and the air smelt fresh and good. We had been in bed but a few minutes, when it started. The whole sky appeared to become a mass of red tracer and searchlights. The Luftwaffe was out to pay its nightly visit to the beaches.

These nightly visits by a handful of enemy aircraft to bomb the shipping off the beaches and the inland concentration areas often hit targets but in the scale of the invasion the Luftwaffe caused relatively little damage. Perhaps the most injury to individuals and equipment was from falling anti-aircraft shell splinters. After their first night ashore, in 159 Brigade 3 Monmouth recorded in their war diary: 'We have already learnt lesson No. 1 after last night. Dig deeper holes.' A 7.2-inch battery belonging to 8 Army Group Royal Artillery (AGRA) proved to be noisy neighbours for 8 RB, and they attracted the attention of the Luftwaffe when night firing.

The 11th Armoured Division's concentration area between the villages of Cully and Lantheuil.

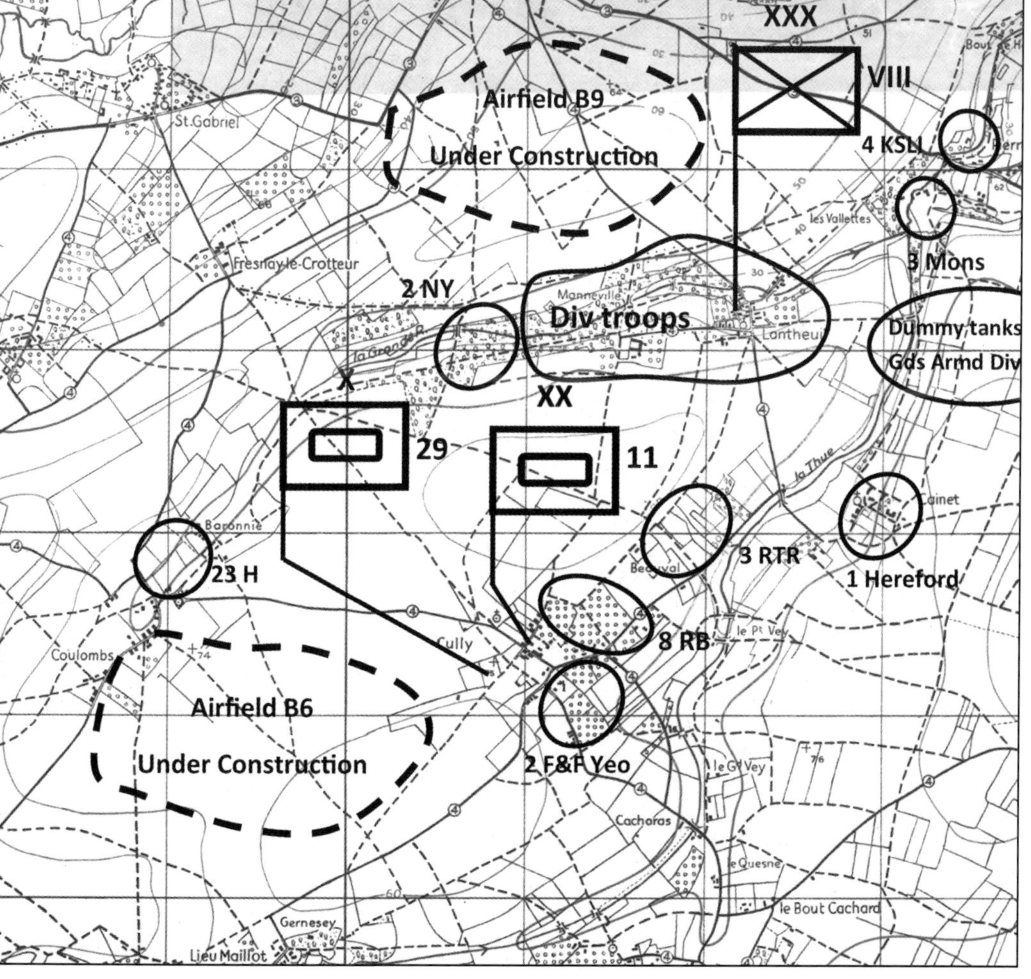

Captain Bell noted:

The following morning we [G Company] had the unusual experience of being roused at 0530 hrs by the Commanding Officer. The night had passed without incident, and it was difficult with the return of the daylight not to believe we were not just on another scheme. First thought was breakfast, for we were very hungry. Happily, we had no need to use the twenty-four hour packs, neither the hexamine cookers. We opened up a [fourteen-man] compo pack and cooked our meal on our No. 2 [pressurised fuel] cooker, a much more civilized affair, with a petrol burner. We even had no need to use our water-purifying tablets, neat as they were, for a good supply of drinking water was found nearby. The final party arrived shortly after we had finished, having spent the night on a landing-craft and reporting it none too pleasant. They had experienced the air activity which we had witnessed. With them they brought the echelon. The whole Battalion was now altogether, harboured in the orchard.

On trips to the surrounding farms, it was found that produce was readily available to supplement compo rations and a vigorous trade in the less popular tinned rations was soon established and some developed a tase for the fiery Calvados apple brandy.

Gunners of the RAF Regiment sited around the nearby temporary airfields were among the anti-aircraft batteries around Cully.

The Pause

General Montgomery always intended that formations, especially those new to battle, would have up to a week to adjust to being in an operational theatre and to ensure that their first battle went as well as possible. However, Hitler had ordered that the Allies be contained in their lodgement, which was just 10 or so miles deep, and with Montgomery being unable to in-load resources sufficient for a second battle while Cherbourg was being captured, there was little scope for early commitment of the 11th Armoured Division. And then, at an early stage in the logistic build-up, a north-easterly storm struck between 19 and 21 June that all but closed the beaches, destroyed one of the Mulberry harbours and badly damaged the other. During this period of bad weather, stocks of combat supplies were badly run down at a point when a significant logistic 'cushion' had not yet been established. For example, in the ammunition supply chain in Normandy the number of 25-pounder artillery rounds was down to a critical level of only twenty-five per gun.[2] In these circumstances, the 11th Armoured Division remained in its concentration areas for almost two weeks. They were, however, far from inactive. The war diary summarised the period 15–25 June:

> Bn. remains in conc area, less 'G' and 'H' Coys who move out to join Armd Regts in their conc areas. Once everyone is organised and ready for battle the time is spent in going to see enemy eqpt, paying visits to other units (a useful way of learning to read French maps) and in organised exercise and rest.[3]

The moves of G Company to join 23 Hussars and H Company to 3 RTR were made on 17 June, but F Company remained in place as they were already adjacent to the Fife & Forfar Yeomanry's harbour area in Cully. E Company's anti-tank guns and mortars also left to join the motor companies. Away with 23 Hussars at Coulombs, Captain Straker recorded:

> In the evening one could go out and there was a small Estaminet nearby where the only drink sold was Calvados, a drink one buys when it is the only one sold. It was remarkable how little damage had been done and it was apparent that there can have been little fighting in the area. We went to look at a few brewed up tanks and anti-tank guns, and there were one or two roadside graves. There was a German radio station which was known by us as the 'Strongpoint'. Major Mackenzie was never happier than when rummaging about this place and bringing back Spandaus and grenades with which the stores and platoon trucks were loaded to capacity, and dozens of bottles of Vichy Water. He also conducted two TEWTs [Tactical Exercises Without Troops] in the pouring rain.

In Major Mackenzie's exercises officers and NCOs discussed tactics for fighting in the parcels of bocage country that could be found surrounding villages and in stream valleys in the area. This was light bocage compare with that encountered later in the campaign during Operation BLUECOAT but the thick-hedged

Major General Roberts (right) with commander 29 Armoured Brigade, Brigadier Rosco Harvey.

banks surrounding paddocks and orchards were both a problem for the attacking infantry and tanks and cover for the enemy. As a result of his visits to other armoured formations, Brigadier Harvey gave a lecture on lessons learned and trips were organised to inspect knocked-out armour. Of these, the Fife & Forfar's war diary noted that: 'As many lessons as possible were learned from units who had been in action and from the siting of positions of various KO'd 88 A.tk guns and Br tks in our neighbourhood.' Meanwhile, in the fields near Cully, G Company and C Squadron 3 RTR were busy. 'All Tk Comds attended demo of action of Tp of Tks and Mot P1 on contact.'[4]

All of this kept those who held rank fully-employed, but for the riflemen like Rolland Jefferson, having removed the last of the waterproofing and repacked their half-tracks so that they could get under the armoured sides, were beginning to wonder:

> For the next ten days we were static and did no fighting and we began to wonder whether we had been forgotten. The thunder of war was ever present in the distance. The whistle of huge shells fired from the battleships out at sea (*Warspite* and *Rodney*) could be heard above us as they fired over our heads into the German positions around Caen. Many of us had all our hair cut off in case of lice infestation. We looked pretty gruesome with no hair.[5]

Preparation for Battle

General Montgomery's next 'colossal crack', Operation EPSOM, was aimed at capturing Caen but it had been delayed from 22 June by the storm and the need to build up sufficient stocks of ammunition. Also delayed was the arrival of VIII Corps' 43rd Wessex Division and 32 Guards Brigade.[6] When the battle began, the tail of the former division and the whole of the latter brigade were still assembling in Normandy.

On 18 June, having abandoned the intent to attack out of the tight airborne lodgement east of the Orne (Operation DREADNAUGHT), according to VIII Corps' historian: 'One day later, on 19th June, a conference was held at the Marie in Creully, presided over by General Dempsey, at which the main intentions of Second Army were revealed for the new phase of operations about to begin.' The plan that emerged over the following days was based on the task laid out in Lieutenant General O'Connor's VIII Corps Operation Order (Op O):

> On 'D' Day (26th June) 8 Corps will break out of the existing bridgehead on the front of 3 Canadian Division with a view to the Corps forcing crossings over:
> (a) The River Odon,
> (b) The River Orne,
> so that at a subsequent date the Corps can be positioned on the high ground north-east of Bretteville-sur-Laize, thereby dominating the exits from Caen to the south.

The enemy facing VIII Corps to the west of Caen were the thinly spread battalions of 26 SS *Panzergrenadier* Regiment, a part of the 12th *Hitlerjugend* SS Panzer Division. The 'fanatical' young SS soldiers had proved themselves to be intractable opponents since their first clashes with the British and Canadians on D+1 and were loathed. Repeatedly ordered to attack, heavy losses eventually forced the SS over to the defensive while panzer divisions, including those of II SS Panzer Corps, were summoned from elsewhere including the Eastern Front. The EPSOM blow would principally fall on just two of the *Hitlerjugend*'s battalions. Against SS infantry who were overextended holding a long front, there was optimism across VIII Corps that the 15th Scottish Division would break

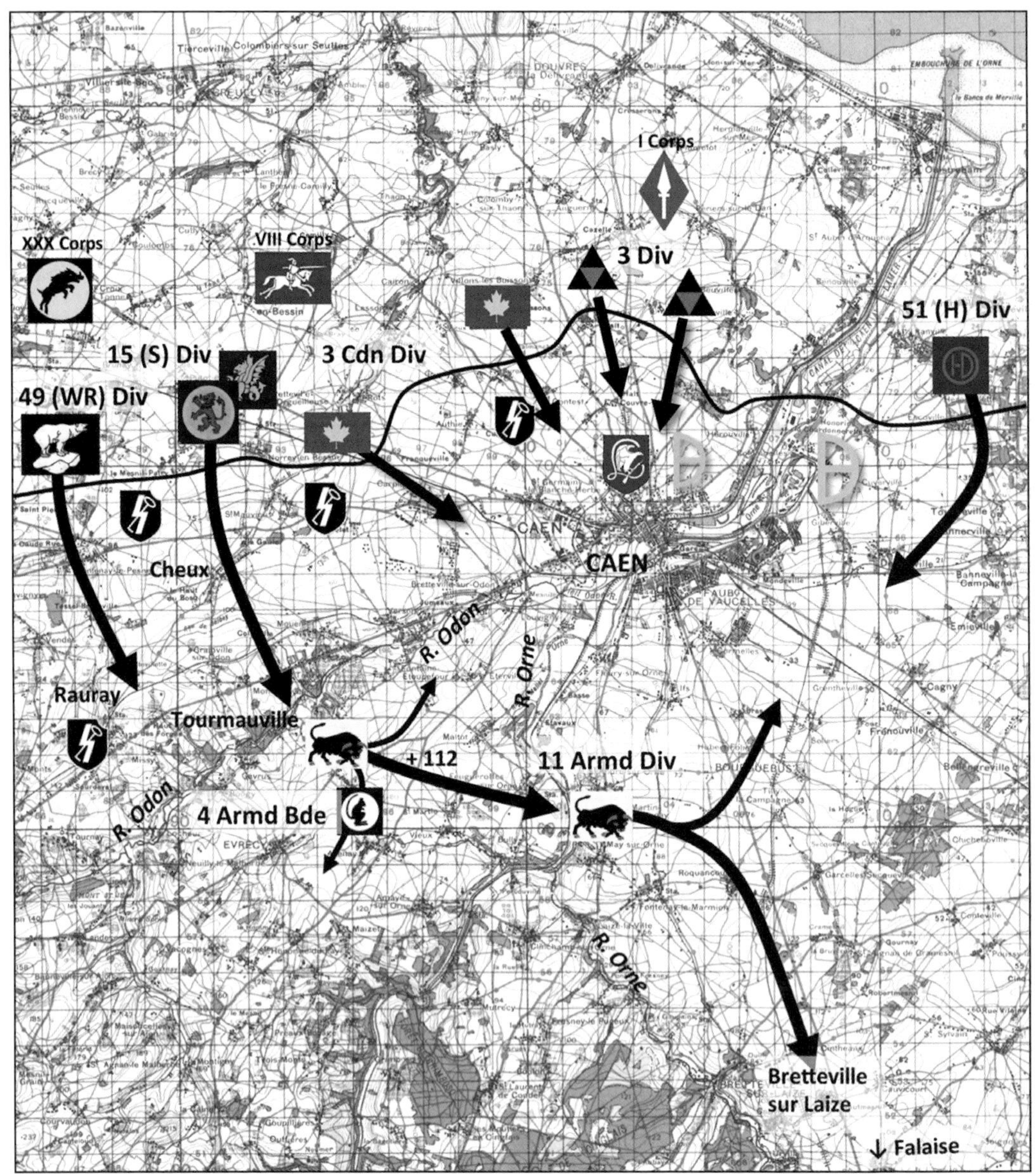

Operation EPSOM was the centrepiece of the offensives to be launched by all three corps in Normandy. Some of the other operations were in the event cancelled or scaled down.

through quickly and reach the Odon some 4½ miles into the enemy position. From there, the armoured regiments and motor companies of 11th Armoured Division were to spearhead the drive to the next river obstacle, the Orne and on to the more open country beyond.

Major General Roberts' divisional Op O followed on 23 June, with the 'Intent' being that once the 15th Scottish Division had secured a bridgehead across the Orne, '11 Armd Div will est itself in the area MAIZET 9557 – ESQUAY and est a bridgehead over the ORNE between 965555 and 005615.' The first phase,

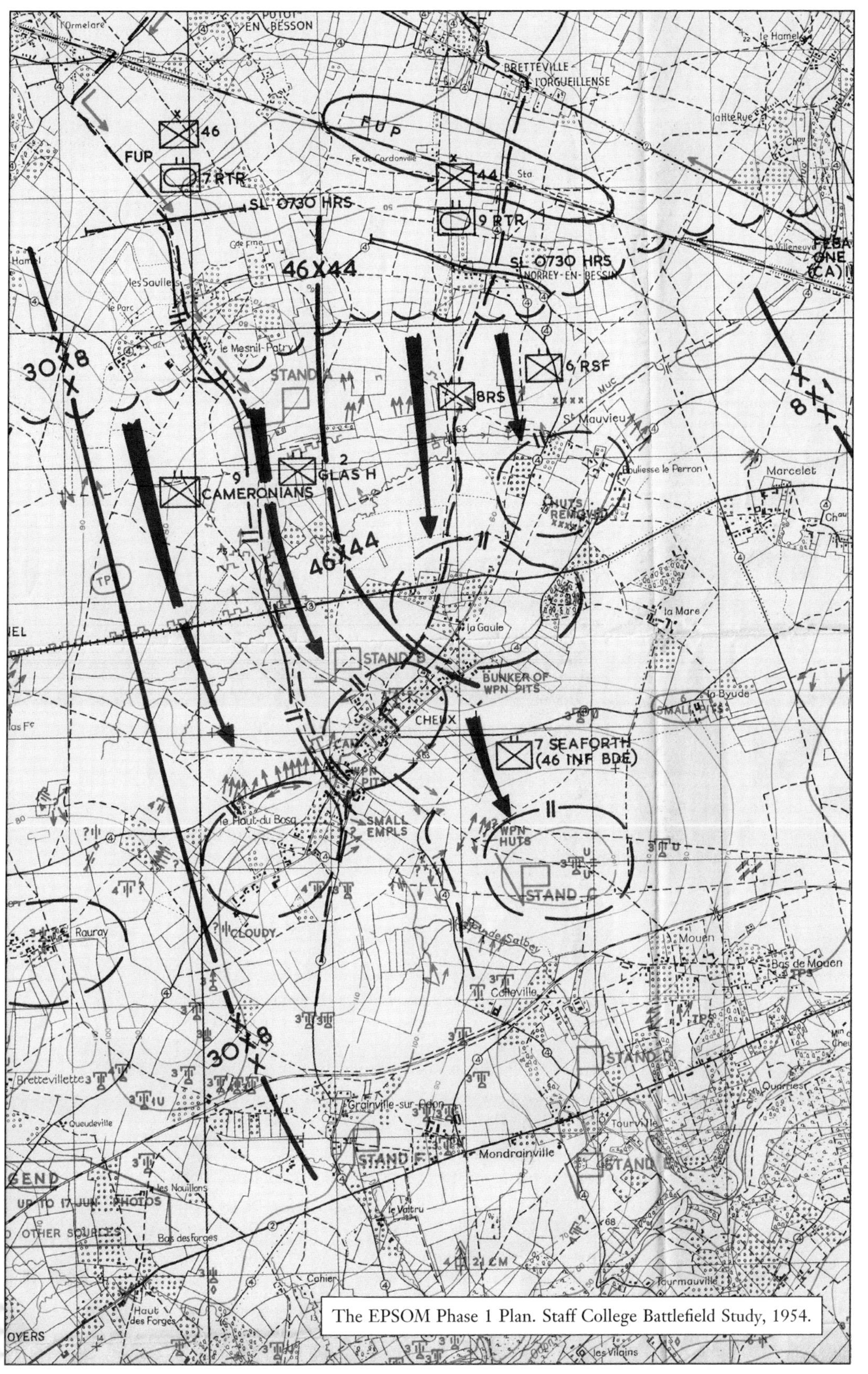

The EPSOM Phase 1 Plan. Staff College Battlefield Study, 1954.

however, proceeded more or less as expected. Initially 29 and 4 armoured brigades, the latter under command of the division, were to follow the advance of the Scots on the cleared and marked routes of MOON and STARS respectively. It was envisaged that the Scots would reach Le Haute du Bosque by 1200 hours and the Odon by 1430 hours.

During 23–25 June conferences and O Groups were held, as orders were formulated and delivered down the chain of command. The division was fortunate that in its first operation, battle procedure was not rushed, nor were there shortcuts. Captain Bell recorded that:

> The following day, Sunday, the 24th of June, brought with its sunshine and warmth our orders at last to move into battle. Whilst the final touches were made to the vehicles and equipment, Order Groups persisted throughout the day. We sat under the apple trees in the welcome shade and heard the great plan unfolded before us. The 15th Scottish Division were to make a breach in the enemy's line in the area of Cheux, some miles to the south, and we were to follow through leaving Carpiquet aerodrome to our left, cross the River Odon, seize two pieces of high ground in our stride – Hill 112 and Hill 113 – and cut the main road running out of Caen to the south-west, after crossing the River Orne. The marked maps were laid out before us. We looked at them and at each other. The operation was certainly ambitious, for it was expected we would cross the Orne the following night, but we, fresh from our training, green, but full of confidence, believed we could do it.

With the sound of fighting growing as the 49th West Riding Division's preliminary battle to the west of EPSOM rumbled on, Captain Straker recalled that Lieutenant Colonel Treneer-Michell joined H Company in Coulombs for a pre-battle pep talk: 'The Commanding Officer came to give us a last address before we set off. He stood on what was a filled-in latrine pit, and as he kept digging into the ground with his heel as he talked, the Company was double expectant.'

Mortar detachment carriers.

Chapter Three

Operation EPSOM

With rain overnight, 26 June dawned dank and chilly, but the country north of the N13 Bayeux–Caen road was alive with activity, as battalions of 15th Scottish Division marched from their assembly areas to their forming up points (FUP), while artillerymen waited alongside over 600 field, medium and heavy guns. The wet weather caused the planned RAF bomber strike to be cancelled but at 0729 hours the artillery opened a heavy bombardment. Back in their harbour area at Cully and Coulombs, 8 RB's motor companies had loaded their half-tracks and were ready to deploy with the armoured regiments, while battalion headquarters shepherded a miscellaneous group of anti-tank gunners and Royal Engineers into order. There was, however, to be no early committal to battle.

Following a squadron of the armoured recce regiment, 2 Northamptonshire Yeomanry (2 NY), 29 Armoured Brigade drove direct from its harbour areas around Cully on to Route MOON at 0730 hours. F Company and 2nd Fife and Forfar Yeomanry (2 F&F Yeo) reached the Caen–Bayeux railway line, where at the head of the column they halted with the rest of the brigade in the order of march 23 H, Brigade HQ, 3 RTR and 8 RB (HQ), strung out behind them, with those at the rear still in their harbours at Cully. Captain Noel Bell provides an account of G Company's first morning 'in battle' alongside 3 RTR:

> We formed up in line along the track with the tanks in front of us, and halted. It was half-past eight, the time laid down when we would start, but there was no sign of movement. After an hour had passed, we climbed out of our vehicles and walked up and down the column or lay on the grass verges under the high trees that flanked both sides of the lane. In the distance a great artillery barrage was going down in the direction of where our objectives lay.

The companies of 8 RB were surrounded by the medium and heavy guns of 8 AGRA. Rifleman Patience of G Company recalled:

> The noise was very loud and we had no ear protection. The guns started in the early morning and the noise was shattering – it went on and on and seemed like hours. Please don't ask how I felt as we waited for the order to move forward because I could not tell you. All I did know was that I was glad I was not on the receiving end of this barrage.

Ahead of 29 Armoured Brigade, the damp morning and the explosion of shells had combined to produce a thick fog, in which two Scottish infantry brigades supported by the Churchills of 31 Tank Brigade became disorientated. To add to the confusion, the accompanying armour ran into unmarked minefields. This plus the resistance of the *Hitlerjugend* soldiers soon had EPSOM's first phase falling behind schedule. It was not until gone 1030 hours that the head of the brigade started to move forward again, but for G Company towards the rear:

> The morning passed, and it was one o'clock before orders were eventually received to proceed. We turned left on to the main road through Cully and forked left at the far end of the village. At the fork the members of the echelon, who were staying in Cully, smiled and waved their farewells and wished us good luck. It was warm now, and the sun was at its height; all sign of the previous night's rain had disappeared in the heat.

The first and second phases of the Operation EPSOM plan.

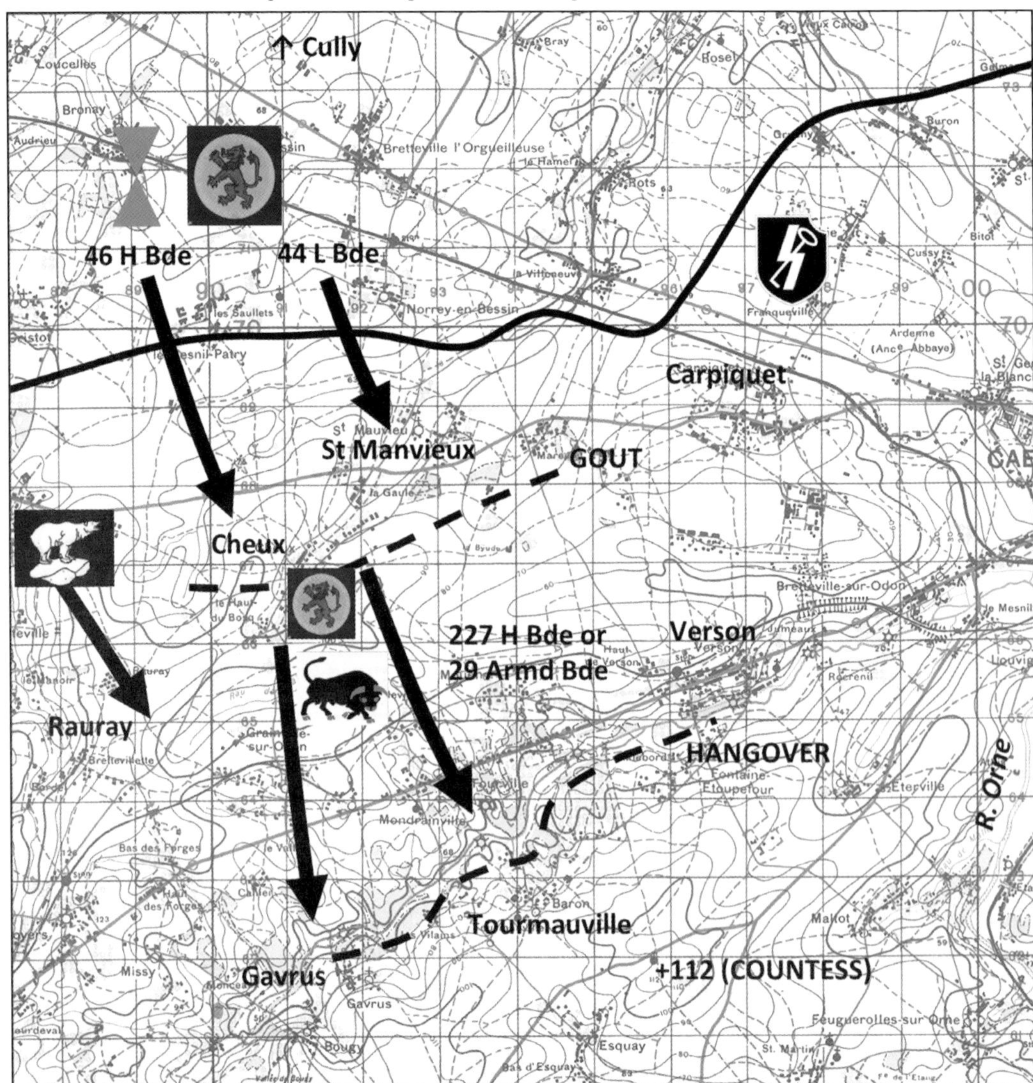

Moving forward only short distances before halting again, G Company had time to absorb the destruction in villages and the fields:

> The roads were bad, being pitted with holes, and we jolted along rather uncomfortably. The country was close for the most part, and thick bushes and high trees frequently hid the surrounding fields from our eyes. We passed ... on to Bretteville-l'Orgueilleuse, where we crossed the main road running from Bayeux to Caen ... [It was] badly mauled, and the ruined houses and shops bore silent witness to the heavy fighting and shelling which had taken place there. A knocked-out German self-propelled assault gun lay between two houses in a little alley-way off the road. It was blackened with fire, and one of its tracks hung limp, shattered and useless across the suspension. Gaily-coloured advertisements for Byrrh and Cinzano, painted on the sides of houses, were spattered with mud from near-by shell bursts ... And all-pervading, above it all, was the sweet, sickly, repulsive smell of death. Outside the village, destruction was no less apparent, and dead cattle, blown and stinking, lay round the smouldering farms. Truly, the four horsemen of the Apocalypse were riding through Normandy.

Meanwhile, Major General Roberts had gone forward, mounted in his Cromwell tank, to where he could read the battle, and be well placed to judge the moment to release 29 Armoured Brigade:

> By 09.00 hours, I felt I had better get a bit nearer the scene of operations, so I got myself into my tank, my ADC in a spare tank and my CRA in another, and set off towards Cheux. We had been going about twenty minutes when, 'Whoomph!' We had come to a grinding halt. We were in a very harmless-looking field, no signs of any kind around and well clear of any battle. We had gone over a mine, part of an unmarked minefield laid by the Canadians, so the Divisional Commander succeeded in becoming the first tank casualty of 11th Armoured Division in the war! No one got hurt, but then a lot of 'prodding' took place. We eventually got ourselves out of the minefield without further casualties, and finally came to rest about 1,000 yards north-west of Cheux, which had just been cleared by the Scotsmen.[1]

Remounted in the spare HQ tank, he was joined by 2 F&F Yeo just north of the village of Cheux, where ahead 227 Highland Brigade and Churchills of 31 Tank Brigade were making slow progress.[2] The Yeomanry's war diary records that their

> ... original role was to push through over the R. ODON and S.E. to the ORNE in support of and co-operating with 15 (S) Div. The Regt advanced behind the inf to CHEUX 9267 where 10 HLI and 2 Argylls [*sic* 2 Gordons] (227 Bde) were pushing thro with some difficulty. 'B' had been leading with the Recce Tp in front but now 'C' Sqn advanced to the S.E. of CHEUX – HAUT DU BOSQUE supported by 'A' on the rt and 'B' in reserve.

Tanks of the 11th Armoured Division's Tactical Headquarters. The GOC's tank followed by the CRAs, an OP Sherman and the escort troop.

10 H.L.I. were to attack thro the centre of village but deviated to the right and their attack did not go fully home. A Churchill Bn was in sp but did not achieve much, following on into the congested village.

With time slipping away, at 1250 hours, Major General Roberts released the Cromwell tanks of A Squadron of 2 NY to the south. Their first problem was to get through to Cheux, where 'the destruction of the village was so great that it took A Squadron a considerable time to find a way through or over the heaps of rubble, shell-holes, and burning buildings'. This was, however, only the beginning of their difficulties, as recorded in their regimental history:

On trying to get out on the far side, they were met by many determined snipers and 'bazooka-men' in the orchards and demolished buildings, behind chicken-houses, high banks and hedges. Several Germans were shot while trying to climb on to tanks with grenades and magnetic mines. Owing to the delay the artillery programme was asked for again and was fired a second time; but even so, A Squadron were unable to make much progress owing to the presence of enemy tanks as well as the previously spotted prepared positions.

These panzers, of 5 and 7 companies, II Battalion 12th SS Panzer Regiment, had been sent east from where they had successfully limited the advance of the 49th Division on to the Rauray spur, to stem the new threat south of Cheux. Claims of knocking out Panthers suggest that they were joined by tanks from the I Battalion.

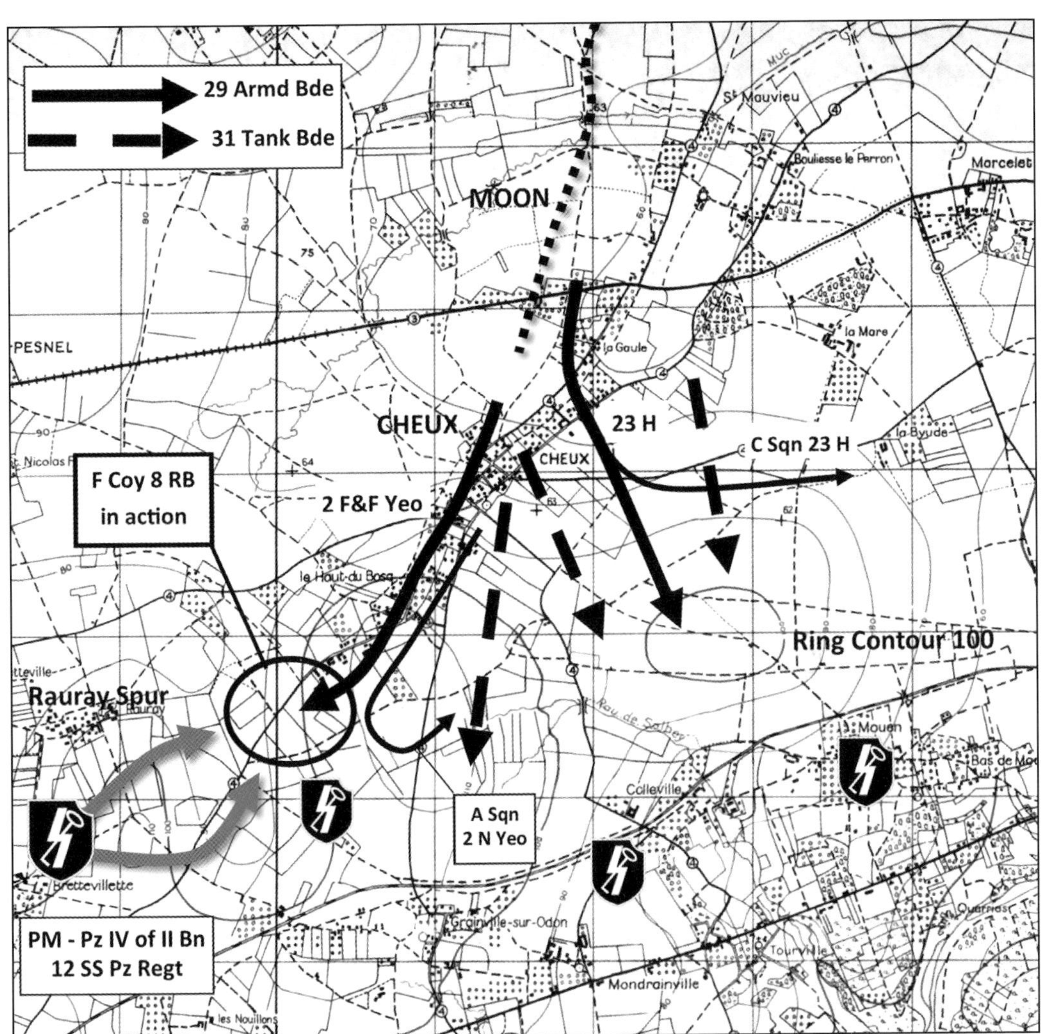

Armoured operations, afternoon of 26 June 1944. The infantry advances by 227 (H) Brigade were coincident with those of the two leading regiments of 29 Armoured Brigade.

Enemy resistance was clearly both determined and in depth, and at 1430 hours, having become trapped in a thickly hedged and embanked field, the decision was made to withdraw A Squadron. Given the next decision, it is worth recording the expected situation in which the release of the armoured brigade and motor companies would be authorised, namely:

> Depending on the str of the enemy in the ODON valley 29 Armd Bde Gp will be prepared to pass through 15 (S) Div at the conclusion of phase II of 15 (S) Div Op or, if opposition is slight will be prepared to cross the R ODON before phase II if the result of the recce by Sqn 2 N Yeo is such that it appears to be possible to seize the crossings with two armd regt gps.[3]

Despite the unpromising results of A Squadron's recce and with the 15th Division having barely started its phase II, the decision was made to release

29 Armoured Brigade. Whether this was made by General Roberts alone or with General O'Connor is not clear, but both wanted to restore the initiative after a slow start by the 15th Scottish Division. Consequently, 29 Armoured Brigade had been committed to battle by 1500 hours. However, in fighting that lasted into the evening, little progress was made and with the infantry of 227 Highland Brigade forward there was no call for the support of 8 RB's motor companies, which remained waiting with the reserve squadrons north of Cheux.

On the two axis that the armoured regiments were to take south of Cheux, the ground is markedly different. To the south-west it is bocage rising up to a ridge bisected by hedges and trees, while to the south-east there was a broad open hill. This is Ring Contour 100, which from its crest stretches down to the Caen–Villers-Bocage railway line, beyond which is 1,000 yards of thick bocage to the Caen–Villers road. A Tiger on this road could for much of its length engage British tanks on the Ring Contour. While 2 F&F Yeo were fighting their way through Le Haute du Bosque into the bocage, 23 Hussars were approaching the open ground and dispatching C Squadron to cover the left flank. Up ahead of A and B squadrons, the Churchills of 31 Tank Brigade were in action and there were already ominous plumes of black smoke from knocked-out vehicles. With the aim of assisting the Gordon Highlanders across the Ring Contour, the railway and into the bocage, the Hussars' Shermans became targets as soon as they reached the crest line. A handful of anti-tank guns backed by equally few Tigers dominated the open ground. With a litter of burnt-out tanks, there was no way across Ring Contour 100 for tanks and only a single infantry company crossed the railway that day.

Waiting to be called forward during 26 June, both G and H companies heard the howl of in-coming *Nebelwerfer* fire for the first time:

> It was a diabolical sound, like that of a giant retching, which was repeated several times in quick succession. This was followed by a great whistling, culminating in a devilish scream, as some unknown missiles hurtled down from the sky. It was the German multiple-barreled rocket mortar, commonly dubbed 'Moaning Minnie'. To those who have never heard one, the attempted description is quite insufficient. The fearful noise as the rockets leave the barrels is beyond true interpretation on paper. Their purpose was dual, combining both psychological and blast effect, and, like the Stuka, the psychological part was frequently more devastating than the subsequent material damage, which was terrible enough.

F Company to the Rescue

On the western flank, having fought their way through Le Haute du Bosq and into the bocage, duelling between the two squadrons of 2 F&F Yeo and the two depleted companies of Panzer IVs was dragging on and little progress was made. Seven tanks were lost by C Squadron and two by A Squadron out on the right, in exchange for claims of two Panthers. With the British advance halted, and the

A tank of 29 Armoured Brigade destroyed by an internal ammunition explosion following a direct hit during 27 June 1944.

7 *Werferbrigade*

The *Nebelwerfer* ('fog' or 'smoke' mortar) was originally conceived in the 1930s to avoid the limitations placed on Germany's post-First World War armed forces by the Treaty of Versailles that limited holdings by the Germans of weapons such as artillery. The name *Nebelwerfer* was designed to obscure the true capability of the system during League of Nations inspections.

The ammunition from the outset included not only smoke but high-explosive and chemical warheads as well. The rocket projectiles were eventually of calibres ranging from 150mm–320mm and were fired from a number of different types of launchers. With thin casings the ammunition, which did not have to survive the muzzle velocities of artillery or mortars, had a consequently large payload for their size.

The presence of 7 *Werferbrigade* under command of I SS Panzer Corps represented a significant asset and narrowed the Allied firepower advantage for the Germans fighting EPSOM. The brigade consisted of two regiments, 83 and 84 *werfer* regiments, both consisting of three battalions.

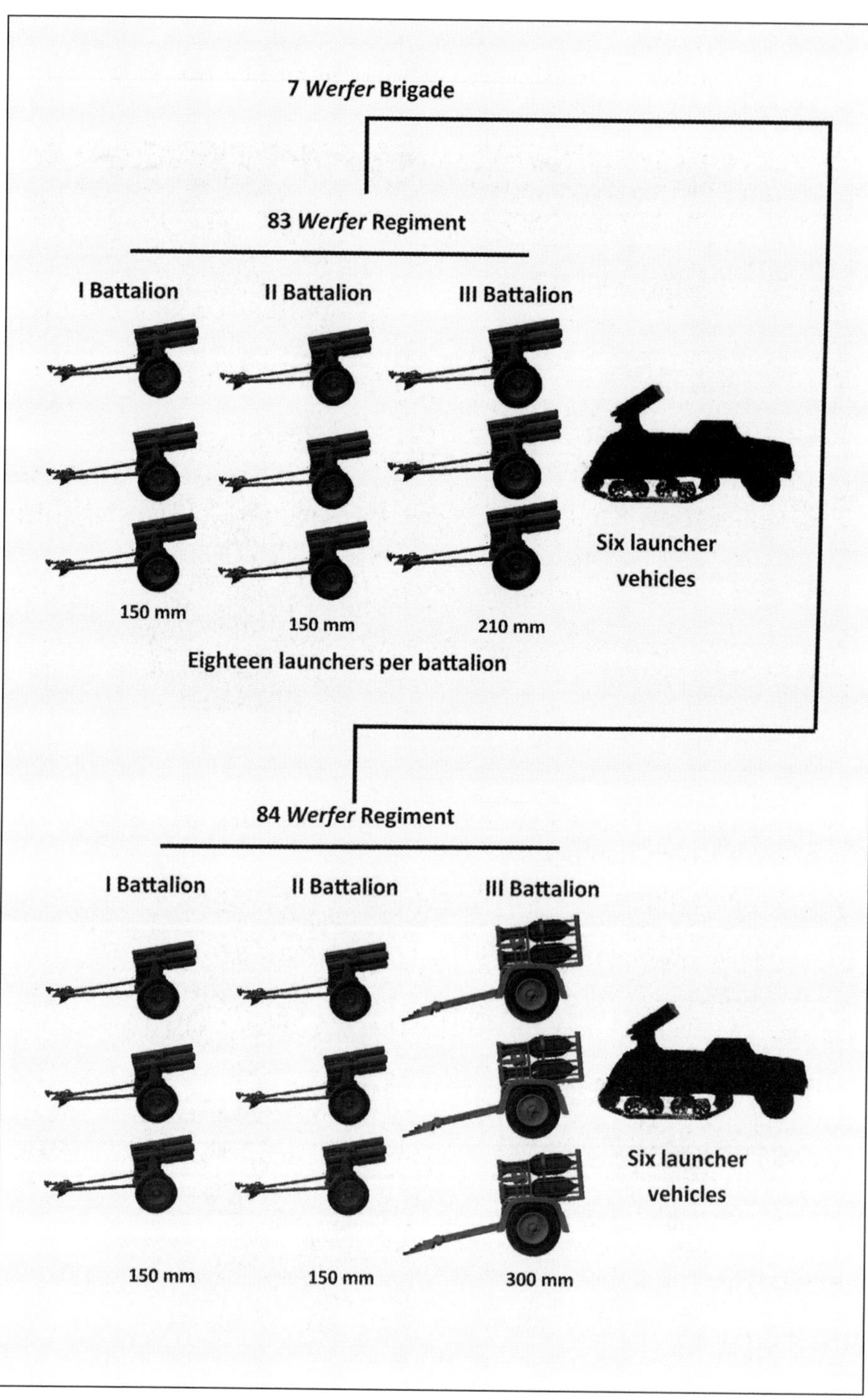
7 *Werfer* Brigade
83 *Werfer* Regiment
I Battalion
II Battalion
III Battalion
Six launcher vehicles
150 mm
150 mm
210 mm
Eighteen launchers per battalion
84 *Werfer* Regiment
I Battalion
II Battalion
III Battalion
Six launcher vehicles
150 mm
150 mm
300 mm

***Nebelwerfer* 41/150mm.** With six rocket tubes, this launcher was the workhorse of the *werfer* artillery. The brigade was established for fifty type 41 launchers. Range 7,500 yards. The HE payload weighed 5.5lb.

A *Nebelwerfer* 41/150mm.

***Nebelwerfer* 42/210mm.** There were eighteen of these five-tube launchers concentrated in the III/83 Regiment. It had a range 8,500 yards and a crew of four men. The HE content was 22lb.

A *Nebelwerfer* 42/210mm and crew deploying the launcher from a hide into a firing position.

***Nebelwerfer* 42/300mm.** The III Battalion, 84th Regiment held eighteen of these six-tube launchers, which had a range of just 5,000 yards and a crew of six men. The HE warhead weighed 99lb.

Two views of the *Nebelwerfer* 42/300mm.

***Panzerwerfer* 42.** Seven launchers each were grouped in I and II battalions of 84th Regiment. They were mounted on the Opel Maultier half-track vehicle, which provided both mobility and protection. With ten 150mm tubes per launcher, the *Panzerwerfer* 42 provided a significant 'shoot and scoot' capability.

A *Panzerwerfer* 42 mounted on an Opel Maultier chassis.

A second *werfer* brigade arrived in Normandy at the end of June and fired in support of II SS Panzer Corps.

SS infantry and pioneers rallied, they began to infiltrate forward along the hedges and ditches to take on the Shermans with *panzerfausts*. At 1815 hours, 10 HLI had come forward but 'after skirmishing' they withdrew 'and consolidated for the night back towards Le Haute du Bosque'. Leaving 2 F&F Yeo without infantry support in the bocage is just one of many examples of poor coordination in their first battle between the Scots and 11th Armoured. To extricate the Yeomanry's tanks, F Company's 7 and 8 platoons were finally called forward but almost immediately suffered casualties:

> As they moved up for this operation, 7 Platoon had the misfortune to suffer a direct hit on one of their Half-Tracks, with the result that they lost ten killed and wounded. Never again in the whole of the Campaign did we suffer so many casualties from a single stonk.

The riflemen of 8 Platoon drove back the *panzerfaust*-wielding *Hitlerjugend* soldiers from the hedgerows, allowing C Squadron in the centre and A Squadron on the right to disengage and pull back. The 25-pounders of 151 (Ayrshire Yeomanry) Field Regiment fired smoke to help the break clean by the tanks and riflemen on top of firing 360 rounds of HE per gun that day. Having disengaged under the cover of smoke, the tanks of 2 F&F Yeo and F Company drove back to the leaguer, passing Cheux in pouring rain, their way illuminated by the depressing sight of burning tanks.

G Company spent most of the evening in a valley as a reserve before moving to join 3 RTR's tanks in the leaguer:

> We stayed there till about midnight, then continued up the slope out of the hollow to form close laager in a field on the right. Most of us slept in our vehicles where we sat, for it had been a long day, and the psychological reaction of some of the things we had seen brought additional weariness.

All three armoured regiments, along with their attached motor companies, leaguered north of Cheux. However, with three divisions all advancing on the same centre line, such were the traffic jams that in the short hours of darkness (2315–0415 hours) only the lucky battlegroups received anything like a full replenishment of combat supplies.

The first day of EPSOM had seen bitter combat, with the *Hitlerjugend*'s thinly spread soldiers fighting with fanaticism. A breakthrough to the Odon had not been made but 15th Scottish Division issued orders by radio at 2300 hours, starting with a review of the situation: 'Enemy now reduced to last reserves on our front. Essential to secure crossings over R ODON as early as possible tomorrow 27 Jun.'

The Second Day of Battle

The morning of 27 June dawned under leaden skies and more rain. The plan was a resumption of the combined efforts to reach the Odon bridges at Tourmauville and Gavrus. However, with three headquarters – 227 Highland Brigade, 31 Tank Brigade and 29 Armoured Brigade – issuing orders, operations were still poorly coordinated. For example, both the 15th Scottish and the 11th Armoured produced their own additional code words and nicknames for use over the radio, not all of which were shared. For 11th Armoured the line of the Odon was nicknamed FORTNUM and for 227 Highland Brigade's bridgehead it was GOITRE. This was an obvious source of confusion and subsequent corps and divisional operation orders specifically banned their subordinate headquarters from adding to the list; a lesson learned.

As the remainder of the brigade prepared to resume the battle to reach the Odon, Lieutenant Colonel Treneer-Michell with the M10 Tank destroyers of 119 Anti-Tank Battery under command took up positions on the right flank to the north-west of Cheux. With the Rauray spur still not in the hands of 49th Division, the flanks of what had already become a narrow salient were exposed and trouble would soon come from the west. Meanwhile, G Company remained with 3 RTR covering the equally open left flank, with 10 Platoon in company with the Recce Troop's Honey light tanks spending most of the morning patrolling out on this flank. As recorded in the regimental war diary, they 'Ran into small A/Tk & Tk screen – engaged them. General line of screen

The M10 Tank destroyers of 119 Anti-Tank Battery normally travelled under command of 8 RB until deployed.

944670-940663-930661. Approx strength 3–6 A/Tk guns, 6-9 Tks.' As the tanks and half-tracks of G Company advanced, Captain Bell recalled the enemy opening fire:

> We rolled down the slope. Suddenly a gun spoke; once, twice, and again. Some sparks seemed to fly off one of the leading tanks, and the air was filled with a sound like that of a racing car passing at great speed – a rushing, whirring note. A moment's pause and the tank burst into a mass of flames. Micky said, 'Eighty-eights.' We were too green to be scared, for we failed to realize the significance of it all. It had not yet registered on our minds that we were in the enemy gunner's sights, and at that moment another armour-piercing shell was being loaded into the breech. It was our first taste of direct enemy action, and it seemed coincidental that the shells were coming our way. We had yet to learn to be afraid. We had yet to learn to respect the German 88 mm. We moved forward a little, then stopped; and moved on again. In such manner, we progressed, while two more tanks fell victims to the guns.

Later in the day, still with the Honeys, 10 Platoon deployed in a screen in front of the main body:

> Some small-arms firing started, and a section of carriers, who were up a track on the far side of the farm with 10 Platoon, brought in two prisoners – our first – having killed a third German who had tried to run away. The prisoners struck us by their extreme youth, looking no more than seventeen. One was a huge fellow with blond hair and blue eyes; a typical Hun. Both appeared intensely arrogant, and we learned they were from an S.S. Hitler Youth battalion.

Meanwhile, as operations were resuming on the right, a 'counter-attack' by tanks from the west was reported by 2 F&F Yeo. What had happened is that overnight a *kampfgruppe* from 2nd Panzer Division had been sent to reinforce the *Hitlerjugend* from the Caumont area. A company of seventeen Panthers reported to the commander of 12th SS Panzer Regiment at Rauray. However, as the situation there was under control at the time, consequently, *Obersturmbannführer* Wünsche, being aware of situation south of Cheux, sent them further east to join his depleted Panzer IV companies. The result was not what has been claimed, a counter-attack, but a failed regrouping, when the poorly briefed Panthers, instead of deploying on the high ground overlooking Colleville, without infantry, blundered into 5th Duke of Cornwall's Light Infantry (DCLI) in Le Haute du Bosque and Cheux from the west.[4] They did, however, add considerably to the confusion in the area. There are numerous claims from various units to have knocked out Panthers, with some almost certainly being hit twice. Colonel Treneer-Michell's M10 anti-tank screen claimed two Panthers before being moved to the Le Valtru area later in the day.

On the right, 10 Highland Light Infantry (10 HLI) and 2 F&F Yeo continued to push south from Le Haute du Bosque towards Granville. They soon came up against the Panzer IVs of *Obersturmführer* Siegel's 8 Company, in well-concealed

Obersturmbannführer Max Wünsche, commander of 12 SS Panzer Regiment.

Obersturmführer Siegel wearing the Knight's Cross he earned during EPSOM.

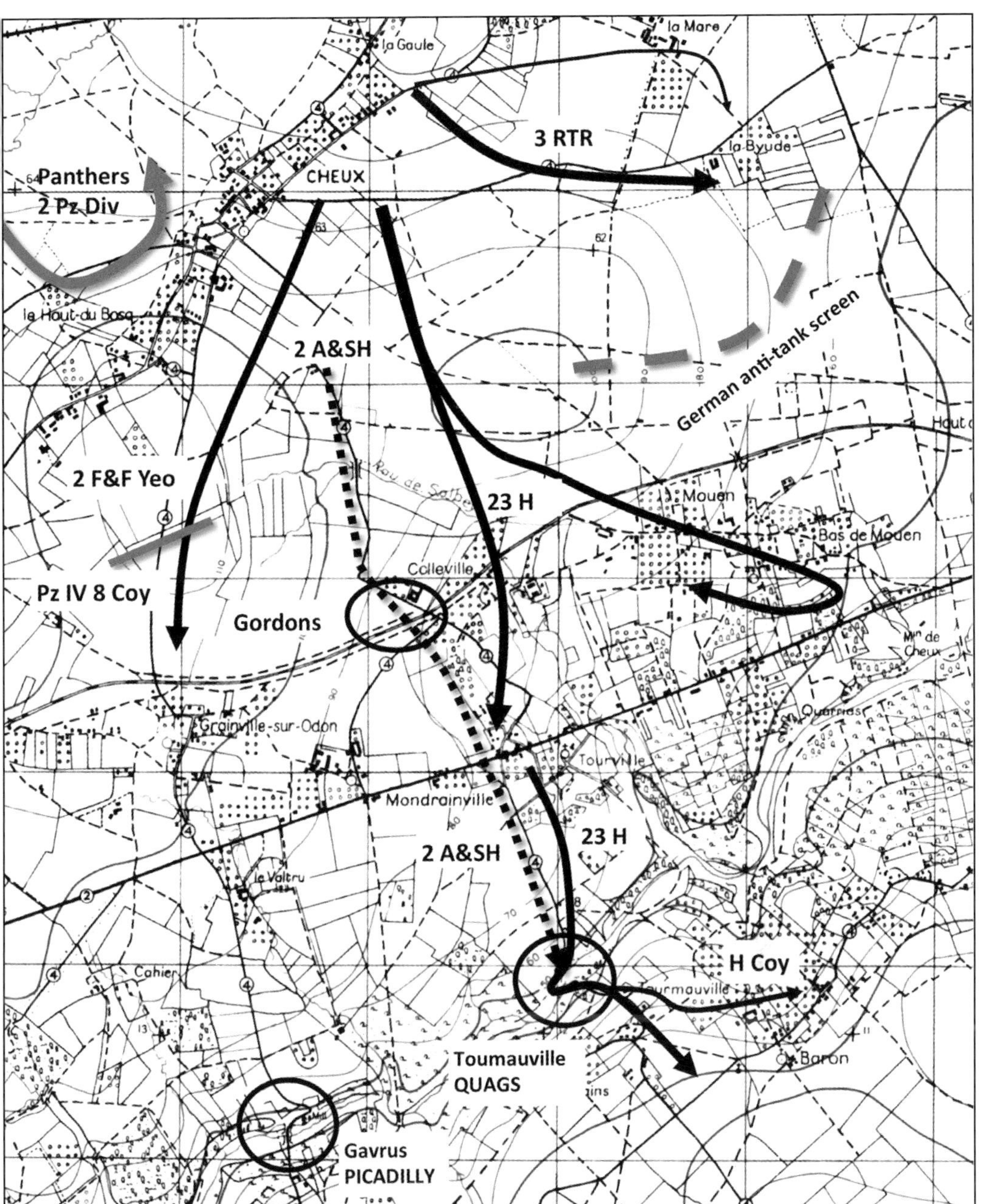

Operations of 29 Armoured Brigade, 27 June 1944.

positions. A combination of main armament HE and armour piercing, plus machine gun fire from the panzers, brought both Scottish units led by A Squadron to a halt. With just four panzers, 8 Company inflicted significant casualties and delay on the advance south from Le Haute du Bosque before withdrawing due to a lack of ammunition. The advance south to Granville could resume, aided by the fact that during the course of 27 June, the 49th Division finally secured a footing on the Rauray spur. However, as 2 F&F Yeo summarised in their war diary:

> The German thrust was continuous from the right all day and tho 'A' Sqn pushed on to the high ground N. of GRAINVILLE 9164 the close country on our immediate rt was never cleared and a small counter-attack was even begun in the failing light. 'A' & 'B' sqns shared the tks we accounted for; 4 Panthers and 1 or 2 Tigers as well as A.Tk and S.P. guns.[5]

In the centre, 23 Hussars and H Company 'shook out from leaguer' and deployed on the lower slopes of Ring Contour 100. According to the company's account, 'Just as we were moving off,' nosing up to the crest, 'we had our first vehicle casualty; a 75mm anti-tank gun made a hole through the [company HQ] scout car, both Rfn Wright and Rodkoff were inside at the time and lucky not to be hurt.'

The German view of the fighting is recorded in the telephone log of the Seventh Army at 1250 hours. 'O.B.[HQ] informs General Hauser of the situation of I SS Panzer Corps, which did not develop as expected, On the whole we have scored a good defensive victory there.'[6] That, however, was about to change.

Eventually overcoming the *Hitlerjugend*'s escort company that had been deployed to fill the gap east of Colleville, 'We moved over the high ground after the tanks and into the bocage.' Here H Company came into its own, moving ahead of the tanks clearing the hedgerows. 'Our first action was an amusing affair; we were to attack a few houses in the thickly wooded and straggling village of Mondrainville.'

> After our 3-inch mortars and several Bren guns had been firing for some time and the leading platoons reached the first house, Sgt. S. Read of 14 platoon was met by a furious Scottish colonel who told him we had been attacking his H. Q. This could not be helped as the situation was rather confused at the time and no-one quite knew who was where, parts of the village still being in German hands. There were odd shots coming from all directions.

This Scottish colonel was almost certainly Colone Tweedie of 2 Argyll and Sutherland Highlanders (2 A&SH), whose battalion had infiltrated into Mondrainville sometime earlier.

The Tourmauville Bridge

Following a frustrating day's waiting during the 26th, 2 A&SH were ordered forward on the second morning of battle but rather than batter their way through on a broad front, they were to slip across the railway and through the Gordons at Colleville. Their objective was the Tourmauville Bridge (QUARGS). By now the *Hitlerjugend* were in trouble; they had suffered significant attrition and were thinly spread. Benefiting from this, the Argylls made their way through the bocage, reaching the dead straight Caen–Villers-Bocage road between Tourville and Mondrainville by early afternoon. Following a fight with panzers and

The Tourmauville Bridge over the narrow but deep River Odon.

armoured cars of 12 SS Recce Battalion, the leading companies were across the road and, led by C Company, advanced down into the steep and narrow Odon valley, through scattered opposition. With the crossing below them, at 1715 hours, C Company attacked and seized the unguarded bridge covered by D Company.

When word came at around 1800 hours that the Argylls had seized the QUARGS intact, the necessity of getting tanks to them was obvious. H Company remounted their half-tracks and prepared to follow the only squadron of 23 Hussars that could be disengaged from around Mondrainville, C Squadron. Getting tanks through a thousand yards of difficult country to the bridge was challenging, as recorded in 23 H's regimental history:

> As soon as it was known that the bridge was intact, 'C' Squadron was ordered to advance followed by the Recce Troop and two tanks of RHQ, while 'A' Squadron were to remain where they were on the railway line until some reconnaissance could be made of the ground south of the Odon. Even now the journey through Mondrainville was not without its excitements. The road was not clear. Beside the tanks of 'B' Squadron which had been damaged or ditched a Panther had been knocked out and lay with its long barrel stretching right across the road. The result was a good deal of congestion. Two Honey tanks were hit and added to the mounting list of derelicts, and had the enemy's shooting been good the number must inevitably have been higher. The command tank itself had been taken on as a

target by the German gunners as it came through Mondrainville and those listening on the Forward Link were amused to hear the Colonel come up on the air in the middle of the battle saying. 'Get behind me, Sixteen Charlie, there's some b****s shooting me up the dock!'

Lieutenant Colonel Harding's tank was subsequently ditched during the steep final descent and, as noted by Captain Straker, 'he finished the day commanding from a Honey [light tank]'.

The Shermans and the motor company's half-tracks went straight across the bridge at about 1900 hours and through the tightly drawn defences of the Argylls, up and out onto more open ground above. Here the tanks deployed in a screen while H Company cleared first Tourmauville and then Baron. They only encountered the odd surprised German. Meanwhile, B Squadron, having had its own adventures, also crossed the Odon to join H Company and C Squadron. By 1925 hours, a vital crossing of the Odon had been captured and was secure as it could be.

As night fell there was no question of the tanks and H Company recrossing to the north of the river to leaguer. The country back to Cheux was far from cleared and leaving the Argylls at the bridge unsupported was unconscionable. Instead, as darkness fell, the two squadrons drove up onto the open ground of Hill 113, where they formed close leaguer protected by H Company:[7]

> It was past midnight when we moved in; it was a lonely little close leaguer as we could hear enemy Tigers [ominously] retiring for the night into Esquay about a mile away, and as we were far in advance of any other troops that night. One or two flares from aeroplanes lit the sky, but we had a quiet night, with only two hours' sleep in cramped positions in our vehicles.

The Tourmauville Bridge had been seized and held twenty-four hours later than planned, but with the delay would the armoured advance to the crossings of the River Orne that generals O'Connor and Roberts had planned be possible? The Germans were vigorously counter-attacking at Colleville and Mondrainville, with 23 Hussars' historian noting: 'Our neck was somewhat stretched and the Germans were making a determined effort to cut it but were paying little attention to the head.' The brigade's other armoured regiments and motor companies were also committed. For example, 3 RTR had been covering the left flank all day but were to be relieved by a squadron of 4 Armoured Brigade at 2100 hours and were to cross the Odon. However, they and G Company needed to replenish before crossing the river. Captain Bell adds the detail:

> Meanwhile, we had received fresh orders, and in the evening we pulled out and withdrew some distance, passing through Divisional H.Q. on the way to the area of St. Mauvieu. It was dark by the time we arrived, and at a cross tracks we met up with the echelon and halted. We were given ten minutes to replenish – a job that usually took half an hour, even in daylight. Petrol and rations had to be drawn, water-cans filled from the trailer, and batteries

One of 29 Armoured Brigade's Shermans knocked out during 27 June 1944.

changed on the wireless sets. During replenishment, orders were given out. Rather an alarming picture was painted of driving down a road, flanked with thick woods, infested with snipers and bazooka men. We were to look out both sides of the truck all the time, as well as front and rear, and to have rifles and grenades ready. Somebody made a joke, and we laughed softly. It was good to laugh again. It seemed so long since we had done so … We moved into the blackness.

However, such were the route finding and traffic difficulties that 3 RTR and G Company gave up and at 0200 hours leaguered near Colleville. Overnight, the infantry battalions of 159 Brigade moved forward on foot to the Odon, to both enhance the defence of the bridgehead and to extend it to the east and west of Tourmauville.

The German Seventh Army that had earlier congratulated itself on the morning's defensive success were now aware of the loss of the Tourmauville Bridge:

2350 hrs. Chef d. G. describes the situation with I SS Panzer Corps to Chef H. Gr. [Chief of Staff Army Group]. In contrast to what seemed appropriate this morning. Chef d. G. is now inclined to answer enemy pressure with more troops. Chef H. Gr. points out that reinforcements have been granted, and that the main object now is to assure that the reinforcements from II SS Panzer Corps arrive in time.

The Bren Gun

The Lewis Gun, the army's first (theoretically) light machine gun, was heavy, prone to frequent stoppages and its barrel could not be changed for sustained fire, which resulted in it overheating. Its replacement, the Bren Gun (Brno-Enfield) was a version of the Czechoslovakian ZGB 33 that had been adopted in the 1930s and, at 19 pounds, was considerably lighter. As was the case with the No. 4 Rifle, it had undergone several changes by 1944.

The Bren was the bipod mounted section light machine gun used by 8 RB's rifle platoons for dismounted action and was also issued to headquarters for local defence. It was crewed by two men; the gunner and the section's second-in-command, who directed its fire and doubled as the loader.

As a section weapon, the Bren's effective range was 600 yards, but it could also be fired from a tripod for greater accuracy and range, with binoculars to spot the fall of shot, or out to tracer round's burnout at somewhat less than 2,000 yards.

The Bren was a gas-operated weapon, which for obvious logistical reasons used the same rimmed .303 ammunition as the No. 4 Rifle and fired at a theoretic cyclic rate of just over 500 rounds per minute. It was fed by thirty-round magazines, which were habitually loaded with twenty-eight rounds due to a weak spring. In comparison with the German MG 42, the Bren's actual rate of fire was considerably less and in a section *vs* section firefight, the enemy had a distinct advantage. To compensate, the British infantry required a numerical advantage and significant artillery and mortar support.

A Bren Gun on a tripod in the high mount position. Beneath it is the magazine box and the essential spare barrel and utilities carrier. When tripod mounted, performance was improved although the Bren's use in the sustained fire role was nowhere near as effective as the Vickers.

Chapter Four

The First Battle of Hill 112

Hill 112 was our introduction into the true hatefulness of war.

[Captain Noel Bell]

The VIII Corps historian outlined the aim for 28 June and described the ground that now faced 29 Armoured Brigade:

> The Corps' intention for the day was still to carry on with the offensive, completing Phase 1 of 'Epsom' and starting Phase 2, the establishment of forces astride enemy communications south-east of Caen, so that this city could be speedily reduced, and thereby the hinge of the German positions in the west destroyed. This involved an advance from the Odon to the Orne of some 3–4 miles across, for a change, open country in the form of an undulating plain. From the confines of the first river line, however, the ground rises quite steeply for a thousand yards, before starting to level off, and results due east of the bridgehead in a flat-topped summit known for ever to Second Army as Hill 112, which despite its seeming insignificance, in reality dominates both the Odon valley and the country to the north. Two miles to the south-west and just north of the small town of Evrecy, a similar knoll called Hill 113, likewise commands the approaches from north and north-east, and from the defence angle forms an effective complement to Hill 112.[1]

After another short night, 'At about 0400 hrs. we moved out of the close leaguer' and deployed on the slopes above the Tourmauville Bridge. As far as H Company were concerned: 'The company went back and spread out along the hedges near Baron, where we had breakfast … The company concentration area was in an orchard at the east end of Baron. The entrance to this was very narrow and several vehicles got stuck.'

Overnight, Brigadier Harvey issued orders for 29 Armoured Brigade to resume the advance to the River Orne (MASON). Shortly after first light the brigade was on the move. The two squadrons of 23 Hussars south of the Odon began the first phase of the advance that would take them to Hill 112 (COUNTESS), while their third squadron crossed the Tourmauville Bridge to join them. At 0430 hours 3 RTR and G Company also followed from their leaguer near Colleville:

> All of a sudden, we were conscious of being shaken. We opened our eyes. It was daybreak, and we had slept for two hours. Trucks were immediately

An F Company carrier belonging to one of the two machine gun platoons photographed later in the campaign.

started and we moved on. Down the road we went, through Colleville and Mondrainville, both of which the infantry had entered before us. The usual scenes met our gaze; dead Germans, dead cattle, wrecked and burning houses, and the usual stench accompanied them.

G Company's Gavrus Patrols

By 0630 hours 3 RTR were reporting that they were across the river and deploying on the lower slopes of Hill 113, holding the right flank of the bridgehead. G Company followed them across the bridge and:

> We climbed up the steep, winding hill the other side. Near the top, some 500 yards from the crest, we stopped, and Company H.Q. pulled into a farmyard on the right-hand side of the road [at the southern end of Tourmauville]. 10 and 11 Platoons were sent out on a patrol to investigate some wooded area which lay about 1,000 yards to our right rear, backing on to the river. The first patrols brought a nil result, and a second effort was ordered, as it was considered certain that there were enemy in that area, and we could not afford to move on with such a threat to our flank.

The seizure of the Tourmauville Bridge late the previous day and the situation north of the river had prevented speedy exploitation of its capture but it also gave the *Hitlerjugend* an opportunity to redeploy. The troops available were few but Headquarters 12th SS Panzer Division rushed both tanks and *panzergrenadiers* to occupy Hill 112 and Fontaine-Étoupefour, with initially only a single depleted panzer company to the Hill 112 area. The necessity of redeploying the panzers, however, finally allowed 49th Division to gain a fuller grip on the high ground around Rauray and Grainville but at Colleville the fighting in the hedgerows along the route to Tourmauville was as bitter as ever. Where there had been only a few rear details south of the river the previous evening, determined *Hitlerjugend* soldiers were now probing the bridgehead west along the wooded valley of the Odon, where G Company were in action:

> Meanwhile, the carriers had taken up positions hull-down behind a bluff, observing the far side of the river to the west of where we had crossed, and our 3-inch mortars moved down in support of the patrols. This time the patrols met trouble. [Lieutenant] David Stileman was leading a section of 11 Platoon down a hedge, the remainder of the platoon being some way behind.

Lieutenant Stileman provides an account of what happened:

> As I was leading a section of my Platoon along a hedge, I came across a wounded German soldier. In perfect English he told me that there was another wounded comrade further up the hedge. I moved forward slowly.

G and H companies in the Odon Bridgehead, morning of 28 June 1944.

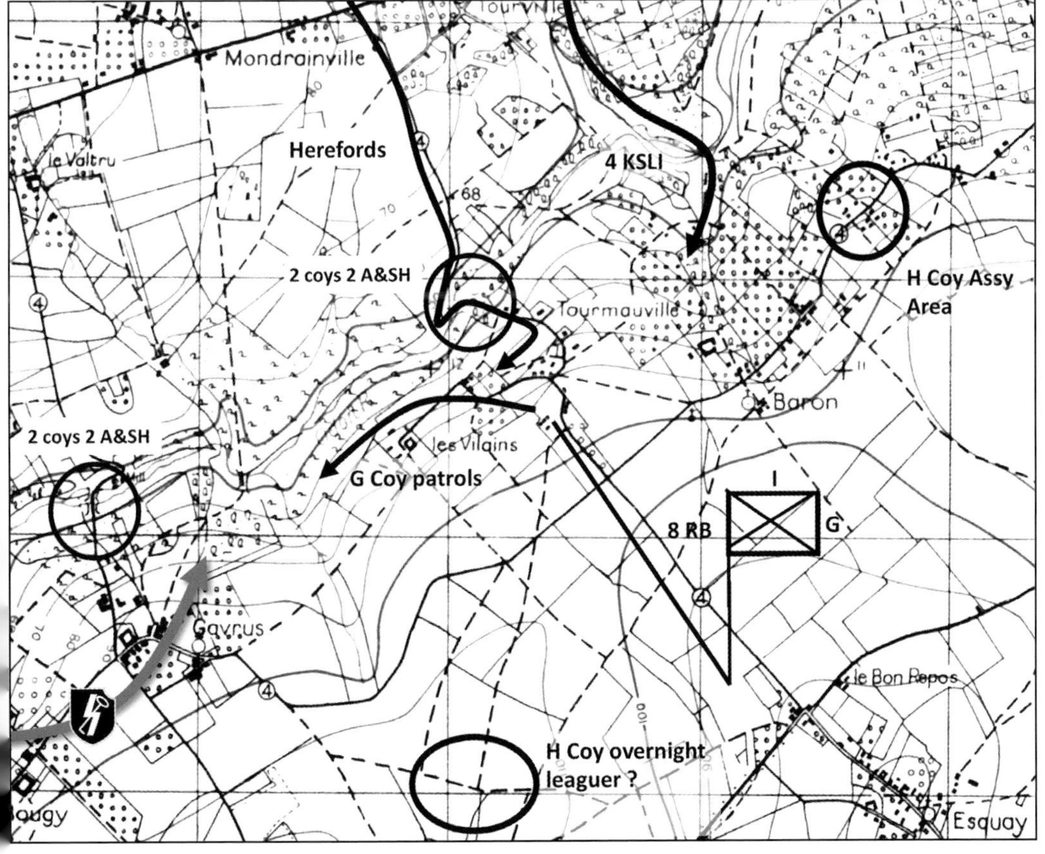

Without any warning, what I now believe to have been a low angled mortar cleverly camouflaged, opened up at virtually point blank range. Instinctively I flung myself on the ground. When I turned round I was horrified to see nearly all the riflemen had been severely wounded. I was not over complimentary to the German soldier I had passed a few minutes earlier. I did my level best to drag Serjeant Deadman to his feet; he couldn't stand … he was paralysed from the waist down. Others in the section sustained gaping wounds from the burst shell or whatever the cursed projectile had been. The remainder of the platoon quite rightly was a tactical bound behind so knew nothing of what had occurred. I struggled again with Deadman but it was hopeless. Feeling wretched I started to retrace my steps when out of the blue Michael Lane who commanded 10 Platoon appeared at my side and wanted to know what was happening.

I lay as flat as I could and described to him where the weapon was concealed. Michael unwisely knelt beside me, searching the hedge through his binoculars. It fired again: I turned to look at Mike, the best part of his throat had been shot away. Blood belched forth from a hold in his chest. He died a few days later.

Rifleman Patience of Lieutenant Lane's 10 Platoon recalled:

My section under Cpl. Peter Bisset was told to attack across a field. We got within about 100 yards when we were spotted. The Spandau, which could file about a thousand rounds a minute, opened up and Peter, who was next about 6 yards away, just dropped down on his face. I knew he was dead. I myself hit the ground and lay there for a few minutes. I looked around and realized that I was alone. If I moved, they would get me, but I couldn't stay there.

There was a ditch either side of me about 25 yards away, that's where the other lads had gone. I remember choosing the ditch on my left. I gathered my thoughts and moved as fast as I could. I got up and ran. The machine guns were behind me and to my right. I was, as far as I can recall, about halfway when they opened fire on me. The bullets went over my shoulders, one went through the side of my beret, then there was a burst about the size of a tennis ball which hit the hedge in front of me. I landed in the ditch and lay still for a while. Someone then spoke to me, it was Butch, our Lance Cpl. He asked if I was all right and I was, but just a few seconds later there was a huge explosion on the edge of the ditch and just above my head. It was a mortar bomb. Three times I had been close to death in a very short time and to this day I still believe that I should have died in that field near the village of Gavrus.[2]

Lieutenant Stileman's platoon and 10 Platoon now commanded by Sergeant Carr were pinned down and needed to be extricated. G Company's history continues:

Micky McCrea went down in the scout car in an effort to get things cleared up. As he approached, he was sniped at, and a bullet entered his knee. He was

Casualty Evacuation and Treatment

The Air Ministry had reluctantly allocated a flight of six Dakota aircraft for casualty evacuation as late as March 1944. The first aeromed flight was on 13 June from Temporary Airfield B3 at St Croix. As the campaign developed, B6 at Coulombs was used, being more convenient to the hospitals around Bayeux.

From the point of wounding, Rifleman Patience was taken to 3 RTR's Regimental Aid Post and evacuated through the medical chain to one of the two Casualty Clearing Stations deployed forward into VIII Corps area for EPSOM. Here he was assessed as being 'fit to travel' and from there was taken by aeromed flight to RAF Wroughton, near Swindon. Having been stabilised, he was transferred by train to a civilian hospital in Birmingham.

In conditions of battle, in order to relieve pressure on the field hospitals, more casualties than normal were evacuated to the UK by sea and air. Rifleman Patience had only the haziest memories of his evacuation:

> I remember being put to bed by some lovely nurses and Red Cross workers. I had nothing, only the dirty clothes I was in, and I hadn't washed for days – I must have looked a right sight and I know I felt it. I woke up, which seemed like hours later, to find two nurses giving me a blanket bath – never had I had one of those before! When they saw I was awake they said 'Hello! How are you feeling?' All I could say was that I was so sorry for the trouble I was putting them to and I was sorry I was so dirty.

Army medics and drivers handing casualties over to the RAF Aeromed crew.

brought back on the scout car, looking very white. The patrol withdrew, and the casualties were evacuated … Meanwhile, some infantry [Herefords of 159 Brigade] had arrived opposite Company H.Q. and were digging in in an orchard on the other side of the road.

To make matters more difficult, I SS Panzer Corps had now begun all-out attacks between Colleville and Mondrainville to cut off the head of the British salient, which while not successful slowed the reinforcement of the Odon bridgehead. Defeating these counter-attacks took a lot of fire support that could otherwise have helped 29 Armoured Brigade on Hill 112. The German troops involved included a newly arrived regimental *kampfgruppe* of the 1st *Leibstandarte* SS Panzer Division.

8 RB was established for six Humber scout cars.

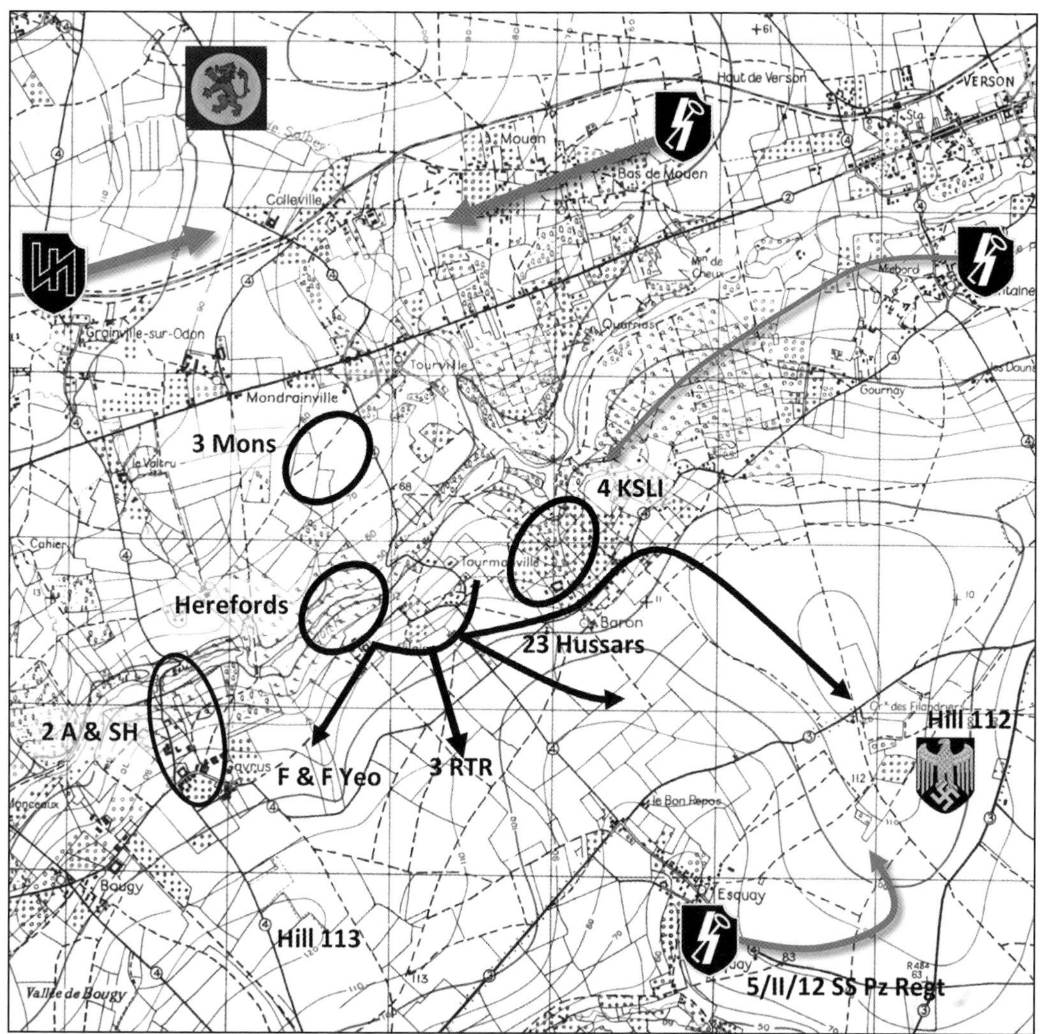

Operations of 29 Armoured Brigade during 28 June 1944.

The Advance to Hill 112

At the same time as 3 RTR and G Company were securing the right flank of the bridgehead, B and C Squadrons of 23 Hussars were preparing for the advance to the Orne 'as soon as it was light enough', but there was to be no swift armoured advance across Hill 112. From their start line near Baron, as daylight came, two Panzer IVs were seen moving around the woods on Hill 112 and were promptly engaged by C Squadron at a range of 1,200 yards. 'One was knocked out and the other hit but managed to limp away.' Captain Straker concluded that: 'This [enemy] position dominated the whole area, and it became clear that we must take this wood before the advance could continue.'

With the enemy alert, at 0530 hours, Colonel Hardie ordered B Squadron to attack around to the left flank, taking a route that followed the southern edge of the villages and paddocks in the dead ground of the Odon valley. C Squadron was to remain in position fixing German attention on Hill 112 and giving covering fire to B Squadron, while A Squadron, when it arrived, would be the reserve.

The flanking move by B Squadron to a position from which they could approach the hill from the north:

> went well for some time over open, undulating country which was good going for tanks, reminiscent of the Yorkshire Wolds. [At about 0600 hours] One tank was hit by a 50-millimetre shot which broke its track, and Lieut. Cochrane's tank was hit and destroyed. The crew got out and came under heavy fire from both sides.[3]

At 0705 hours Major Seymour reported that B Squadron was 'nearing COUNTESS'. They had driven as far east as the Roman road that crossed Hill 112 north to south, where they deployed to advance up to the crest.

However, on reaching a distinct rim where the broad plateau of Hill 112 opened out, despite superior numbers, B Squadron was brought to a halt:

> The enemy opposition in the area consisted of dug-in tanks and infantry in position in a small wood. Their tanks had alternative sites to move to under cover and were almost impossible to get at. An attempt was made to knock them out with some self-propelled guns which were under our command and were sent forward with 'B' Squadron. It was unsuccessful. Medium artillery was tried without effect.

The Woods of Hill 112

There is considerable scope for confusion over the various pieces of wood on the plateau of Hill 112, with the various adjutants/intelligence officers using a variety of names for them when compiling war diaries and then not even consistently! Working out which 'wood' accounts are referring to is not always straightforward, often requiring reference to several sources to triangulate an answer.

The 'wood' most often referred to was in fact an orchard, referred to henceforth as 'the Orchard', but, being surrounded by a hedge of tall trees, it looked like a wood from the northern rim of the plateau. The Orchard was also surrounded by a bank and divided into two by a ditch, bank and a hedge, but this latter feature was soon shredded by shell and mortar fire.

The second wooded feature, referred to in this book as 'Quarry Wood', was in fact a relatively small chalk quarry again surrounded by mature trees.

To the south of Hill 112 along the road to Avenay was a copse used by the Germans for tactical headquarters and assembly.

The final wood was a small orchard located between the main Orchard and the Croix de Filandrières. This orchard was quickly shredded by artillery fire and became an insignificant feature that is rarely mentioned in accounts.

All three woods provided cover in an otherwise open country, but it was the Orchard on the crest of 112 banks and cover that was the focus of the fighting during the battles of Hill 112.

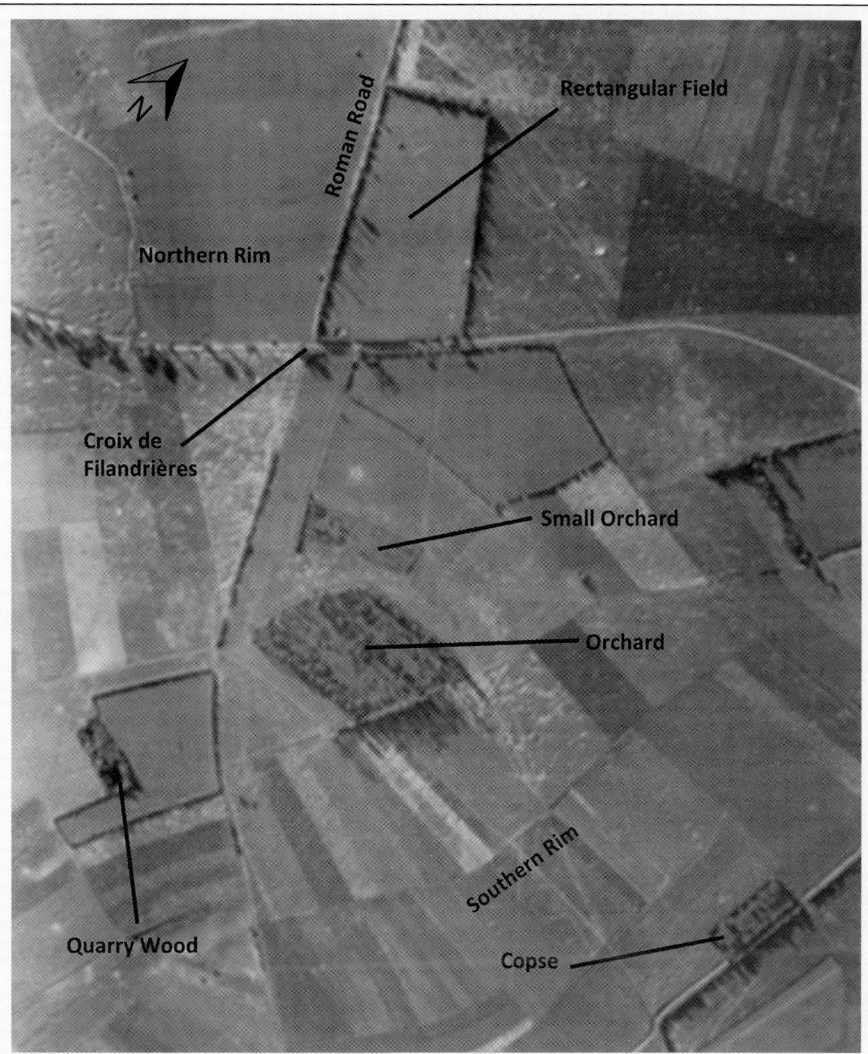

An aerial photograph of the plateau of Hill 112.

Captain Straker of H Company noted:

> This wood was marked on the [50,000] map, but was not the same shape as
> was given, and however many times one walked round it, one could never
> quite get the lie of the land in one's head. In addition, there was beside it a tiny
> orchard, which complicated the position.[4]

Visitors should note that the gap between the Orchard and Quarry Wood has been
reduced significantly by the planting of a memorial wood around the base of the
modern electricity pylon.

The Defenders

The German tanks on Hill 112 at this hour were identified by the British as Tigers. Two weak companies of Tigers were known to be deployed against the Scottish Corridor. *Sturmbannführer* Hubert Meyer, *Hitlerjugend*'s chief of staff, cautioned against the Allied identification of Tigers on Hill 112 early on the 28th:[5]

> It must be noted, regarding these observations, that Tigers were most probably not in action on Hill 112. Those attached to 'HJ' Division fought in the Verson area. In the context of the battles on Hill 112 it is worth noting that the tank unit 23rd Hussars was facing the attacks by parts of *Panzer-regiment* 12 before noon and in the afternoon.

The German armour known to have arrived in Esquay at the foot of Hill 112 at approximately 0700 hours that morning were the ten Panzer IVs of 5 Company, II Battalion, 12 SS Panzer Regiment. One of the panzer platoon commanders, *Obersturmführer* Kandler, stated that they had broken contact near Cheux and redeployed south of the Odon during the night, their exhausted crews like those

Sturmbannführer Meyer, the 1a or chief of staff of the 12th *Hitlerjugend* SS Panzer Division.

of their opponents getting very little sleep. Other Panzer IVs of *Sturmbannführer* Prinz's II Battalion also followed across the Odon during the morning of the 28th, leaving 8 Company to block the advance south of le Haute du Bosque.

The presence of Tigers later in the day is more credible, with 101 *Schwere* Panzer Battalion's war diary recording: '*Hauptsturmführer* Mobius knocks out six enemy tanks and is then himself knocked out. The remaining Tigers block further British advances at Verson & Hill 112.' The emphasis in the diary is on blocking rather than counter-attacking.

There is also evidence that guns of a Luftwaffe motorised battery of 1st Battalion, 53 *Flaksturm* Regiment, were deployed in the immediate area, with its 88mm guns in the ground role.[6]

A German general standing on the crest of 112 and looking at the commanding views is reputed to have been heard to say, 'He who holds Hill 112, holds Normandy'. This may or may not be true but what is certain is that the defences on the hill were prepared during the spring of 1944 as a part of the plans to deny Caen to the Allies. The SS panzer crews certainly would not have had time to dig their tanks in during the night of the 27th–28th, let alone alternative positions.

Colonel Hardy followed B Squadron in his by now un-ditched HQ tank, but it was proving very difficult to locate the enemy:

When he arrived on the hill, he could not find a place from which he could see what was going on, so he moved round to a position [further west] which not only gave a good view of the hill but overlooked the village of Esquay where the enemy, obviously very surprised, could be seen running around …
As an observation point the position was excellent, but though it offered

A Tiger emerging from cover.

a perfect view of this stretch of country, it was quite without cover and obviously no place to stay in the circumstance. As RHQ moved away the CO's tank was hit.

The CO survived, remounted and concluded that much of the fire was coming from the flanks of the hill.

H Company in Action

Colonel Hardy's recce confirmed that the first step to resume the advance across Hill 112 was to establish control of the woods on the crest of the feature and came to the same conclusion as the German general; it was 'vital ground' that dominated the area.[7] With tank losses mounting, at about 0900 hours, Major Mackenzie received orders for H Company to join the attack on the crest of Hill 112. The company was still waiting in their assembly area at the eastern end of Baron, from where they drove east to the Roman road and up through the corn, debussing from their half-tracks below the Rectangular Field and the tanks on the plateau's northern rim. The two artillery observers with 23 Hussars, one from 13 RHA, the brigade's own regiment, and one from the supporting medium regiment, both mounted in OP tanks, produced a fire plan, incorporating H Company's mortars, which included smoke. Direct fire on enemy positions would be provided by the Shermans of B Squadron, which had by now been joined by C Squadron to their left.

With the woods of Hill 112 some 500 yards ahead wreathed in the smoke of exploding shells, H Company advanced over the rim, across the Caen–Évrecy road and on to the open plateau with 14 and 15 Platoons leading followed by company headquarters. The reserve was 16 Platoon along with one of the carrier sections, while the other carriers covered the left flank.

A report of the British advance up Hill 112 was passed to 7 *Werferbrigade* by field telephone from an OP on the northern rim of the feature. *Feldwebel* Doorn, who answered the telephone, ran to get *Hauptmann* Gengl, 6 Battery's commander, who received the following message:

> Sir, the British are on top of the hill. A Sherman tank has stopped just 5 metres from one of our observation posts. For God's sake don't ring – they'll hear it. We'll try and get back somehow. I don't know what has happened to *Leutnant* Wernike and *Leutnant* Nitschmann. I think they must have been overrun.

The two officers had indeed been captured, by 14 Platoon with the aid of German grenade collected at Coulombs. One rifleman recalled:

> Suddenly a scraggy-looking beggar in field grey appeared from a hedge hatless and with his hands in his air. He was rushed off at the point of a bayonet. He kept looking back, frightened or perhaps worried about what was happening to his companions. A moment later, two more, one an officer, were captured.

The view across the plateau of Hill 112 from the Croix de Filandrières to the woods and Orchard on the crest.

Climbing the bank and breaking through the hedge that was already shredded by shellfire, they entered the Orchard and drove the few assorted Germans troops back. The riflemen followed up across the central ditch and hedge, through the blown-down apple trees to the far hedge, where they started the reorganisation phase. This included platoon and section commanders arranging a hasty defence by tying up arcs of fire, while the platoon sergeant redistributed ammunition and arranged the evacuation of casualties to the company aid post. During this time they were joined by the company's support weapons and carriers.

Meanwhile, *Feldwebel* Doorn had been dispatched with a patrol up the open southern slopes of the hill, in order to confirm what was happening on the crest, but he was driven off, losing two men. His report to *Hauptmann* Gengl confirmed: 'It's not just a couple of tanks up there! Tommy's got anti-tank guns and part of a machine gun unit.'

In line with German tactical doctrine, an immediate counter-attack was ordered. Two companies of *Sturmbannführer* Prinz's II Battalion along with a company of Panthers that had only just arrived below Hill 112 went straight into action, with only the briefest orders being necessary. Spread out across the hillside, the panzers advanced at speed. *Oberscharführer* Kretzschmar, commanding a panzer, recalled:

> After a very short assembly, we started the attack in a broad wedge formation on the wooded area. We worked our way forward, each panzer giving the

Oberscharführer Kretzschmar's Panzer IV and crew.

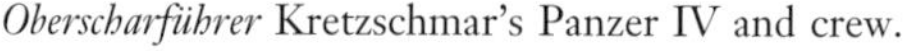

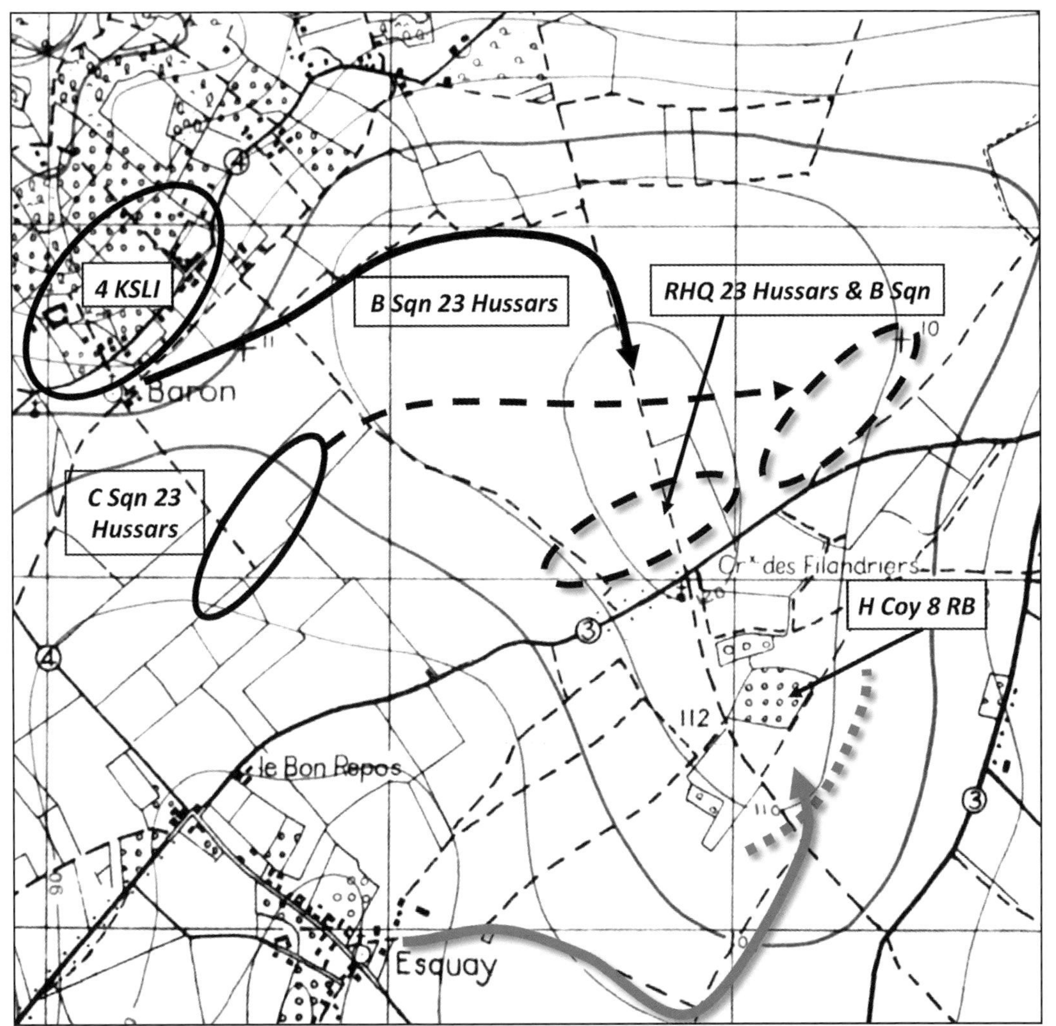

The deployment of 23 Hussars and 5th Panzer Company's counter-attack on Hill 112, morning of 28 June 1944.

other covering fire. Without firm targets, we fired anti-tank and explosive shells into the wood. The attack moved forward briskly. When we had approached to within 300 to 400 metres, we spotted English soldiers between the trees. We fired the turret and forward machine guns into the wood.

Kretzschmar's platoon commander, *Untersturmführer* Kändler, recalled that: 'We detoured around a small wood approximately three quarters up the hill and approached the square wooded area in a wide loop without being bothered much. I had come within 100 metres of it when I suddenly spotted the brown uniforms of Tommies rushing to and fro.' Kretzschmar adds more detail:

At approximately 100 metres from the wood, we changed from the wedge into a staggered line since the gap between the two woods was only 80 to 100 metres wide . . . I was now driving as the point panzer. Our direction was approximately northwest … We cautiously made our way forward along the small [Quarry] wood which was 150 to 200 metres wide. Behind me drove the vehicle of *Unterscharführer* Jürgens. His gun was pointing in the direction of the [Quarry] wood so as not to be surprised by a shell from an enemy 'stove pipe' [bazooka]. At the end of the wood, I ordered an observation halt and searched the ground in front of us with my binoculars for tanks and Pak [anti-tank guns]. Since I did not spot anything suspicious, I ordered 'Panzer march!' After a drive of only 10 to 15 metres there was a sudden bang, sparks were flying and we noticed a hit from the right, 3 o'clock [right] direction. I shouted at the driver, *Sturman* Schneider: 'Reverse, march!' He reacted at lightning speed, threw the panzer into reverse, and backed into the cover of the [Quary] wood at full throttle. Not a second too soon, otherwise the British would have nailed us directly. Immediately in front of our bow, anti-tank shells ripped ugly black furrows into the green grass.

Captain Straker, on the receiving end of the counter-attack, recalled:

At the far end of the wood [Orchard] three enemy tanks counter-attacked, firing both HE and MG at very close range. The platoons began to move back through the wood, at the same time a friendly tank put the whole of Cpl Rick's section in 16 Platoon out of action; meanwhile the enemy tanks caused many casualties especially in 13 [Scout] Platoon.

Without *panzergrenadiers* accompanying them and with stout trees in the hedge and enemy infantry at close quarters, armed with anti-tank weapons, the panzers did not press home their attack into the Orchard. However, as Captain Straker recorded, such was the volume of fire from the tanks that 14 and 15 platoons were driven back to the cover of the central hedge and ditch, where they took up fire positions to resume the fight, as described by *Untersturmführer* Kändler:

Now, the wood came to life. Fire from rifles and machine guns was pinging against the armour, we were covered by mortar and artillery fire. We did not hold back either and briskly returned the fire as we were backing away. We returned to our assembly area without any losses. There, we inspected our damages – a clean hit, gone through between the engine and fighting compartments approximately 25cm below the turret. Except for a small shrapnel stuck in my right thigh, we all escaped with just a scare.

Sturmbannführer Prinz rallied his panzers for another attack but with the guns of VIII Corps now on call as a priority, the Royal Artillery observation officers called for a YOKE target and broke up the attack before it closed on the Orchard.[8] Kändler concluded: 'We came under violent fire, had losses, and were pushed back under constant fire from armour-piercing weapons to the base of Hill 112, behind a hedge.' From then onwards small numbers of panzers worked their way up to the plateau's southern rim, from where they joined mortars and *nebelwerfers* in keeping the Orchard under a steady main armament and machine gun fire.

Untersturmführer Willi Kändler (left) and *Oberscharführer* Willy Kretzschmar.

Close Air Support

During the battle, 8 RB and the armoured regiments of the brigade, as recorded in the headquarters war diary, had their first experience of close air support by aircraft of the Second Tactical Airforce operating from Advanced Landing Grounds (ALG) in Normandy. This included the orchard on the crest of Hill 112.

> During 28 and 29 Jun use has been made of Typhoon rocket firing aircraft to shift dug-in Tiger and Panther tks. Sp was called for through a VCP [Visual Control Party] with Tac HQ netted into Bde Comd net, aircraft and airfield. Targets were indicated by Red smoke fired by RHA. Calls were generally answered within 30 mins and although not always accurate, the appearance of aircraft had a good morale effect on our own tps.[9]

Procedures in the early days did not always go to plan, as this example from the Hussars' account of the fighting on Hill 112 testifies:

> The Gunners put down red smoke to indicate the target. One round fell amongst our own tanks and the hillside was immediately covered in yellow smoke, tins of which were issued to each tank so that it could signal to our aircraft and assure them that it was friendly. It often worked. On this occasion the CO dropped the smoke in the turret of his tank, to the great amusement of those who were near enough to see what happened and the discomfiture of his crew, who found it rather overpowering.

VCPs were joint RAF/army parties that typically worked at brigade level. Alongside a gunner officer, the key member was a pilot who could talk to the aircraft in terms that would be understood by the other pilot up in the attacking aircraft. They often had to go forward into the battle area to perform their function.

G Company Joins the Fray

Meanwhile, 2 F&F Yeo and F Company arrived south of the Odon and took over responsibility for the right flank from 3 RTR, who were ordered to deploy to the left flank where it seemed that enemy activity in Fontaine-Étoupefour threatened trouble. The fighting there, however, remained in the Odon valley and the decision was made that the RTR would relieve 23 Hussars in place in the centre. Following the morning's fighting, the Hussars were now very short of ammunition and with the intensity of the fighting it would have been impossible for the trucks of their A Echelon to complete a 'replen' anywhere on Hill 112.

G Company moved with the RTR's tanks and took up position in the Rectangular Field, not forward in the Orchard from which H Company were to be withdrawn:

It was early afternoon now. We climbed aboard our vehicles and, describing a circle in the field in front of us, emerged at the top of the hill by a cross-roads. There we passed over diagonally, and continued across country. We were heading due east now, and the village of Baron lay on our left. There was a scrap going on there between the K.S.L.I. and some enemy infantry. As we drew level to the village we swung up right and, leaving an L-shaped wood to our left, started to climb the lower slopes of Hill 112.

We had been told to expect trouble here, for there were a lot of Boche reported in the thick corn which grew waist high on the slopes, and also in the woods to our left. We ascended with the half-tracks in column and the carriers guarding the flanks. We rumbled over the fields, slowly climbing. The carriers were firing their Brownings into the woods, and suddenly return fire started. Bullets pinged off the sides of the vehicles with monotonous regularity. We made sure our heads were below the level of the armour-plating, while the drivers adjusted their visor-slits to the minimum. Throughout, we continued climbing slowly, so slowly it seemed, up the hill. As we neared the top. Typhoons swept down and fired their rockets into a small wood, by the summit, sending up great columns of dirt. The small-arms firing ceased, and we turned into a square field which was surrounded by tall trees and bushes. There was only one gap, and a steepish bank had to be negotiated before entrance could be gained. The Company aligned itself along the sides of the field, and the anti-tank guns were put into position. This was Hill 112.

Meanwhile, forward in the Orchard, H Company was preparing to abandon their half-dug fire trenches. Captain Straker wrote:

Eventually we were ordered to withdraw some time before 1400 hrs. The vehicles went up to a field just behind the wood, 16 platoon tracks and a few vehicles from the anti-tank platoon being set aside for casualties. There were no stretcher-bearers, and it was most uncomfortable for the wounded, who were carried out of the wood and lifted on to the vehicles; this was done under occasional and light fire from the enemy tanks who were firing through the trees and high hedges around the field. We then drove back in a column to the dead ground of the cornfield behind the road, and were ordered back to concentrate round a farmyard at Tourmauville, a village about one mile west of Baron.

On the way, we passed the 2nd Fife and Forfar RAP [at the bottom end of the Rectangular Field], where we left our wounded; most of the company then lost the way and it was evening before we found the right place [Tourmauville].

The battalion's war diary recorded that H Company had suffered about twenty casualties on Hill 112. Captain Straker concluded that: 'This was a tragic day, the casualties were all men who had trained and lived with us in England and everyone was rather shattered at their loss.'

It took the Germans some time to realise that the orchard had been abandoned by the British and to give orders to reoccupy it and for other panzers to work their way into positions on the southern rim. In the meantime, G Company began digging in in the rocky limestone soil of the Rectangular Field. 'Nearly all the half-tracks were ordered to withdraw, as their presence was redundant and, apart from giving away our position unnecessarily, they were easy targets to hit.' Having dropped off ammunition and digging tools, the trucks returned some 600 yards down into the Odon valley and leagued behind the RTR's reserve squadron. Meanwhie, up in the Rectangular Field, after half an hour of digging, G Company came under machine gun fire, 'somewhere from the direction of the wood that the

SS troops manning an MG 42 in the light role.

Typhoons had previously attacked'. As more Germans occupied the Orchard and positions on the far edge of the plateau, 'for some while spasmodic firing continued'.

> It then became apparent that the tempo of the fire was increasing, and soon a great crescendo was reached. The tanks, who were on the left of the field, hull-down behind the ridge, returned the fire, for the Company from their position could not see whence it came. For several minutes the sky was rent with a great flow of tracer in both directions.

Rifleman Habertin recalls that, after being in the Rectangular Field for some time:

> The storm broke. The enemy had been watching us settle down and before a single trench had been dug, down came those dreaded 'moaning minnies'. There was nothing to do but lie down and bite the earth. A half-track a few yards away went up in flames and when the mortaring finally stopped, the complete battalion was in a state of utter chaos. All the company vehicles were mixed up, no one knew where their section or platoon was, wounded men were yelling for help and nobody in authority could get any orders carried out.

This bombardment was the harbinger of a counter-attack, this time by the *Hitlerjugend*'s *panzergrenadiers* that had arrived with further Panzer IVs and Panthers. *Untersturmführer* Kändler of 5 Panzer Company recounted that:

> At around 17.00 hours, our total number, through newly arrived Panzers, was back to nine. The 6. *Kompanie* under *Untersturmführer* Helmut Buchwald with *Zügführer* (platoon leader) Kurt Mühlhaus was also there. This time we tried to attack the hill by swinging wide to the left [west] around the small wood in front of the square wooded [Orchard] area. Despite being covered from view to the right toward the square wood, we again came under heavy fire from tanks, had a few Panzer losses and had to withdraw again to our starting positions.

G Company's account reads:

> After a few minutes infantry were seen advancing from the west. They were making full use of cover, and it was difficult to assess their strength. Artillery support was called, and a heartening barrage of 25-pounders showered down on them. Nothing more was seen of the infantry, and there was no counterattack. It was, however, apparent that the position was quite untenable, in view of our strength, through the hours of darkness, and it was decided to withdraw from the summit. Accordingly, the necessary orders were given and the Company, amid a great fresh outburst of shelling, began the descent.

With the tanks of 3 RTR deployed along the northern rim of the plateau, their right flank was vulnerable and they went back to leaguer on the lower slopes,

Sherman tanks of 29 Armoured Brigade in the cover of hedges north of COUNTESS.

leaving G Company isolated in the Rectangular Field. However, other plans were afoot.

During the day, E Company and the Shermans were able to resume the advance south towards Hill 113 and, as recorded in the battalion's war diary.

> 28 1230 WD 'F' Coy with 2 F&F Yeo cross ODON and take up posn on right flank on high ground SOUTH WEST of TOURMAUVILLE, 8 Pl clearing wood in area LES VILAINS 928624 takes 15 PW, identifications 141 Flak Regt and 12 SS Arty Regt.

Out on the open slopes of Hill 113 infantry would have been very vulnerable and of little use with the tanks and panzers duelling with each other from hull-down positions. F Company commented that they remained around a farm near Baron for the rest of the day and the following morning, being shelled and mortared. Overnight 8 Platoon dispatched tank-hunting patrols but the panzers had withdrawn or were out of range for the platoon's PIAT.

Night Attack

VIII Corps at this stage issued orders to pause the advance to the Orne but for the 11th Armoured Division to 'consolidate current positions'. This, of course, for 29 Armoured Brigade meant Hill 112. Rather than trying to hold positions on the crest with a single motor company, 8 RB was to conduct a night attack to capture and hold the feature. An entry in the battalion's war diary timed 1930 hours reads: 'Orders received for Motor Coys to revert to comd and the Bn. to put in a night attack on the wood 962617. C.O. and Coy Comds go uphill on recce and Bn is ordered to conc 954635 [a paddock at the foot of the Roman road].' H Company, harboured with the 23 Hussars at Tourmauville:

> … reorganised, had something to eat and replenished with petrol and ammunition. This was made all the more difficult by repeated orders to move at once and by machine gun fire from tanks of another of our armoured brigades, who were firing for no reason at all into our harbour from about 2,000 yards away to the north, on the high ground back over the river.
>
> When we did move, the driver of one of the leading carriers was killed by this fire, causing the whole column to stop and making the day seem even more gloomy.
>
> There was rubble across the street of Baron, which made the way hard to find in the fading light, and after much difficulty we arrived at the Bn. RV, which was in the same long sloping cornfield just SW of Baron. As we drove in, enemy tanks attacked firing tracer; all the battalion vehicles were moving, there were several bangs, presumably mortar bombs landing in the area and a half-track which was blazing fiercely.

This half-track at the foot of Hill 112 was used as a landmark by the companies throughout the battle. With orders given and preparations well under way, at 2145 hours the diarist noted:

> Just as 'G' Coy is coming back to conc with the rest of the Bn. an enemy force incl. Tigers appears over the hill from the SOUTH, causing some confusion. 'G' Coy suffers some cas. In moving to assembly area incl. Lieut-Oxley-Boyle wounded, and the Tigers came up on top of the ridge and shelled the Bn. Conc area at 954635.

An hour later 8 RB's attack was cancelled and the motor companies returned to leaguer with their armoured regiments. Having already been engaged by panzers that had seen the withdrawal and then deployed forward to the northern rim, the prospect of a single battalion attacking 1½ miles over open ground, made clear as day by enemy illuminating rounds and against clearly strong defences, was one factor but there was another. Wiser counsel had prevailed. H Company's account explains:

> We stayed in this field for an hour or so, and then the attack being cancelled, we reverted to the 23 H. We were to occupy the same [Rectangular] field for

One of 8 RB's half-tracks photographed in leaguer at the foot of Hill 112.

the night, from which we had dropped the casualties earlier in the day, and attack our old friend the wood [Orchard] early the next morning. The company went up on foot and the vehicles were to follow and form a leaguer in the dead ground just behind the field. It was a very dark night and the vehicles were being controlled by two people, with the result that the two columns got lost only a few yards from each other. At 0200 hrs, though, everyone was in their proper place, and the enemy tanks considerately withdrew for the night.

As ever for the infantry, there was little rest, and patrols were tasked from G and H companies. One by Lieutenant Sudlow and Corporal Hone's section of 15 Platoon was

> sent out on a patrol by the Brigadier. They were to see what was going on in Esquay; they were told they 'need not be offensive, but if they felt like it they could drop a grenade or two into the turret of the odd Tiger tank.'

The patrol brought back information on Tigers in Esquay, while Lieutenant Stileman led an H Company patrol to the Orchard, where he heard the sound of digging. For the *Hitlerjugend*, *Untersturmführer* Kändler remembered that: 'In the late evening hours, our regimental commander [*Obersturmführer* Max] Wünsche drove up to our hedge and praised our action [in containing the enemy]. He told us that thirty-six enemy tanks had been counted on the hill.'

The Battalion in Command at Hill 112

By the morning of 29 June, it was now clear that the II SS Panzer Corps, which ULTRA had been tracking across Europe from the Eastern Front, was arriving in Normandy and preparing to attack the Epsom salient from the west. It was confirmed that with only 29 Armoured Brigade south of the Odon, 11th Armoured Division was to consolidate its position. As far as Hill 112 was concerned, 3 RTR would return to the northern rim, with Colonel Treneer-Michel and his headquarters commanding operations on the hill itself including reestablishing positions in the Orchard. An entry in 8 RB's war diary at 0800 hours sets the tone for the day:

> 'H' and 'G' Coys revert to Bn. comd. 'H' Coy pushes on again to the wood with 2 pls of 'G' Coy while the rest of 'G' Coy takes over the field and Bn HQ Gp. (less 119 Bty) moves up to area 958624. This posn was held all day without charge, being all the time under observed fire from 88mm's in the area CARPIQUET 9767, as well as mortars from the EAST, SOUTH and WEST.

The day had started well for G Company in leaguer at the foot of the hill but, as recorded by Captain Bell:

> Then, like a bolt from the blue, came the order that the Company would return to 112. A man nearby said softly: 'Oh Christ!' It reflected the feelings of us all. The order had been given, and it was not 'for us to reason why.' We prepared ourselves and by mid-morning we were ready to start. We climbed on to vehicles, and began the ascent once again.

H Company, who led the attack provides an insight into the plan:

> At 0800 a very sudden attack was ordered on the wood. We had one platoon of G Coy, under command, for this attack and later, for consolidating we were given another G Coy. motor platoon. We also had an anti-tank and machine gun platoon from E Coy, for consolidation. The carriers had been left for the night in the cornfield area by Baron, Sgt. Millwood's section coming up as soon as we were established in the wood.

Standartenführer Kurt Meyer, commander of the *Hitlerjugend*, watched the bombardment from the ridge above Caen:

> … a massive barrage hit Hill 112. Would the British anticipate our plans and attack before we did? With an uncomfortable feeling, I watched tanks of 11th Armoured Division climb the slope south of the Odon and take Hill 112 in a pincer movement. The summit could no longer be identified. The impact of heavy gunfire was tossing the Norman soil around, metre by metre. There was no longer any doubt. The British had launched a pre-emptive attack.

H Company's attack went in and 'almost immediately enemy mortar and shellfire fell on the Orchard, which was itself clear of any enemy'. The previous evening

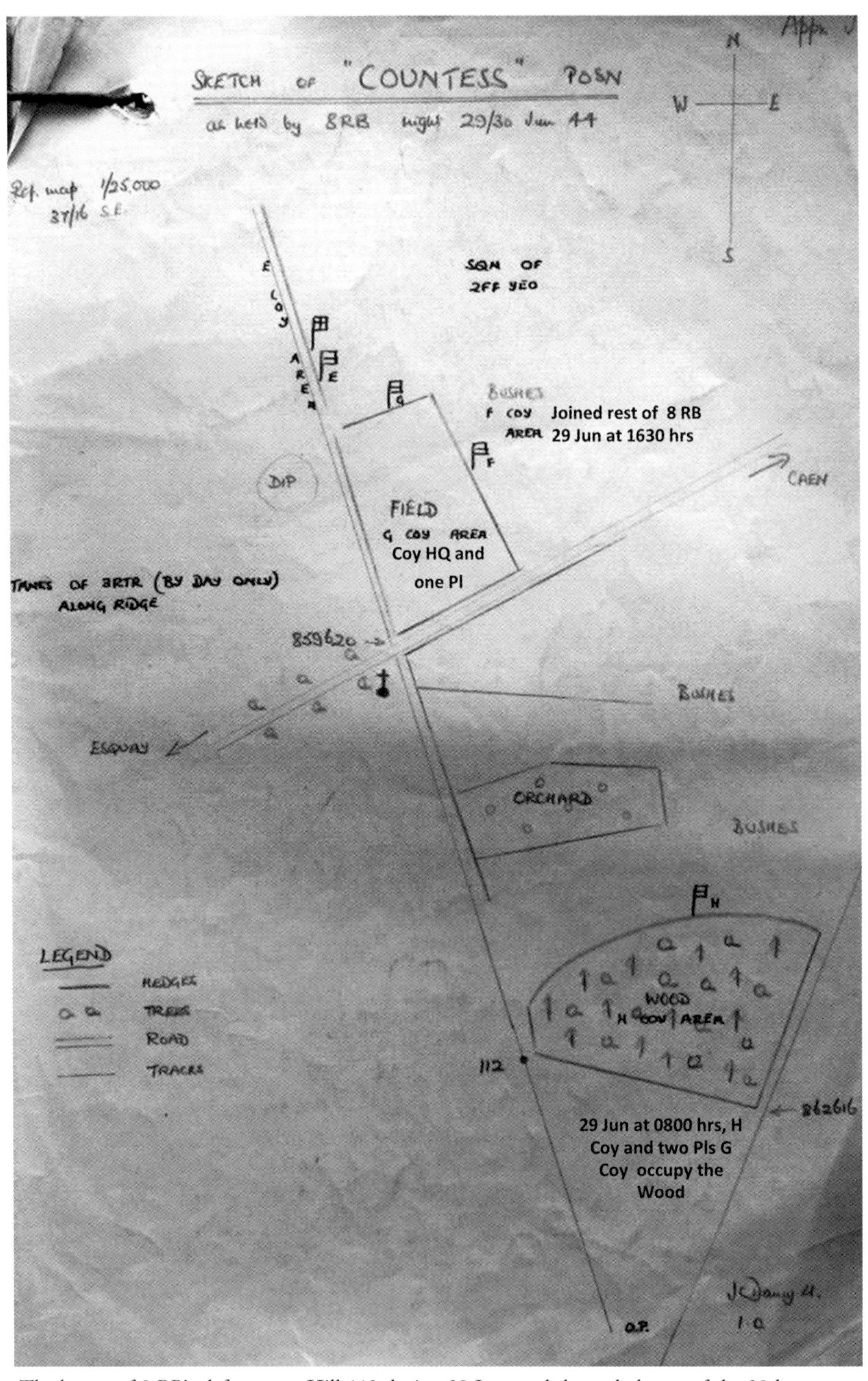

The layout of 8 RB's defences on Hill 112 during 29 June and the early hours of the 30th.

the Orchard had been heavily defended but why was it abandoned? The answer was that there had been a very serious failure of liaison between the *Hitlerjugend* in I SS Panzer Corps and the attackers from the west, II SS Panzer Corps. *Sturmbannführer* Meyer, the chief of staff, had not been advised by his counterpart in the 10th *Hohenstaufen* SS Panzer Division of the mounting delays II Corps was suffering in launching their attack.[10] Having withdrawn from the area before dawn, all that could be done was to use as much of the scarce artillery and mortar ammunition as possible to hit 8 RB as they sheltered in their shallow trenches of the previous day:

> Major Mackenzie was wounded during this and was helped back by some members of 14 Platoon, some of whom were wounded on the way. This was a bad blow for the Company, for he had commanded us for over three years and had been splendid in the wood on the day before.

Meanwhile, E Company and Battalion Headquarters established themselves in the Rectangular Field below the plateau's northern rim. They might have been out of direct fire, but one casualty was a dispatch rider's motorcycle. Rifleman Ayre was probably one of the last DRs to be de-horsed in the companies. Other motorcycles had already been damaged or abandoned and, with there being no replacements, the riflemen were re-employed elsewhere in the battalion.

As casualties mounted, 13 Platoon's carriers were taking back the wounded and bringing up ammunition. One passenger on the way up was the reassuring presence of Lieutenant Colonel Treneer-Michell, who soon arrived in the Orchard. He walked around braving enemy fire to visit the riflemen in their trenches and 'explained to everyone why it was so important for us to sit in this position, under observed enemy fire from all directions'.

Meanwhile, back on the northern rim of the plateau, to improve the situation to the west, Colonel Silvertop ordered Major Close's A Squadron 3 RTR to take and hold Esquay. The squadron's point troop, however, suffered the same fate as 23 Hussars' CO the previous day, losing two tanks as they crested the ridge. Only a little later, at 1010 hours, the code word LIMEJUICE was flashed to the RAF officer at brigade headquarters, requesting an air strike on a dozen Tigers and Panthers and six self-propelled guns. This resulted in a series of attacks by Typhoon aircraft that neutralised the enemy.

Having returned to his headquarters in the Rectangular Field, Colonel Treneer-Michell requested reinforcement of the Orchard. The RTR sent forward a troop of tanks to provide close support from the cover of the hedges, but this was not popular, as recorded by Captain Straker:

> They insisted on having some 3 RTR tanks in the wood, which drew fire on us. On several occasions 01 [company command half-track] and other vehicles came up to us to evacuate casualties, or to bring us food. This also drew fire, but was worth it … although we had two OP officers with us from the RHA and from a Medium Regiment, we never located any positions,

A 'cab-rank' of RAF Typhoons. Operational analysis report on the results of rocket attacks on ground targets concluded that under combat conditions, the 50 per cent zone for the rockets was 75 yards. That meant that the chances of an eight-rocket salvo securing a single hit on a tank within an area of 200 sq ft, was about 0.7 per cent.[11] Nonetheless, as testified by German soldiers, the effect was terrifying and caused significant damage to unarmoured targets in the area.

and no one would give us an air OP which alone might have been able to spot them.

Conditions for G Company headquarters and 11 Platoon were scarcely better back in the Rectangular Field. 'All day a continual barrage came down, varying in pitch from time to time. Several times the carriers were blown bodily off the ground by the blast, but, luckily, no direct hits were scored.' Leaving company headquarters with a manpack radio, the company second in command, Captain Bell, took the half-tracks back down the ridge towards the outskirts of Baron, in the vain hope of cover:

Down at the hedge, we parked the vehicles. Some of 'H' Company's trucks were already there, and they were running a shuttle service to the top and down again fetching their wounded … A great hail of 'Minnies' came over, and one fell just behind one of the trucks. Shrapnel tore a great gash through the armour plate, as though it were tissue paper, and some ammunition in the back began to ignite. It was only small arms, and we were able to put it out before the bombs which were stacked on the other side of the truck were affected. More mortaring followed and it was decided to move the transport to the western end of the field, about a quarter of a mile away. This was done but it was in vain. We had no sooner arrived when the 'Minnies' came over again, falling right in our new position, damaging two more half-tracks. It was apparent, then, that we were under observation and further movement in the immediate area would be useless. Beside where we were parked was a deep sunken lane. We climbed through the hedge, and took the best cover that we could down there, taking with us an extension lead from the wireless of 01.

Up on the top of Hill 112, G Company were doing their best to improve the defensive layout of the wood area:

All day long we had waited with every nerve alert for the expected counter-attack, while shells and 'Minnies' rained down on us. No. 3 Section of 9 Platoon had been moved to the right flank of the position, and was forward, observing the village of Esquay beneath them. Later, about six o'clock, 10 and 12 platoons advanced over the crest of the hill, and took up positions on the forward slope [hedge], and dug in. Shells, however, exploding in the trees above them evinced the futility of slit-trenches without coverings, as the splinters fell straight into them. In consequence, 10 Platoon pushed out, to the right and 12 Platoon to the left, until they were in the open ground on either side of the wood. There they began to dig in again.

Shells and mortar bombs bursting in the tops of the trees that grew in the hedges multiplied the effect of shelling and produced nasty wounds full of organic material. This also reduced the trees to skeletons, which after their own experience of suffering British shellfire in the Orchard, the Germans nicknamed it the '*Dornenkrone*', or 'the Crown of Thorns'.

At the end of the fighting the trees in the surrounding hedge were shredded. Another German nickname was the 'wood of half trees'.

Eventually, with the enemy still active, Colonel Treneer-Michell sought further reinforcement and at 1630 hours, when 2 F&F Yeo deployed on the left flank of Hill 112, the war diary records: 'F Coy reverts to comd and takes up posn in area 959623 [see sketch map on page 78].' With F Company having joined 'the rest of the Battalion on the since notorious Hill 112' they started dug in to the left of the Rectangular Field, from which direction trouble was expected:

There was quite considerable shelling and mortaring by the enemy, who appeared to resent our presence on this very commanding piece of ground. A recce patrol was teed up from 5 Platoon to go to the village of Esquay, about 2 miles to the south, but two minutes before zero-hour … the patrol was cancelled. The six members of this patrol, frustrated and disappointed, therefore spent the remainder of the night on Hill 112 in their patrolling dress, which proved inadequate against both the elements and the enemy. Shirt sleeves, PT shoes and no steel helmets between them made the night hideous.

Riflemen Girling (left) and Lance Corporal Ellis of F Company photographed during 29 June sheltering behind their half-track, opening a Fourteen-Man Compo box.

Withdrawal

The much-delayed attack by II SS Panzer Corps eventually began at 1430 hours but the two SS Panzer divisions advancing from the west astride the Odon, although having initial success, were halted well short of Hill 112 by 1900 hours. This was, however, not apparent to the Army Commander, General Dempsey, at the time, who anyway:

> believing these attacks to be rehearsals for a more coordinated effort on the following day, confirmed the Corps Commander's appreciation of the general situation and ordered 8 Corps to consolidate in its present area. In addition, in view of the exposed position of 29 Armoured Brigade, he instructed that it was to be withdrawn north of the River Odon. The bridgehead was not, however, to be given up but indeed maintained at all costs by the infantry holding it, 159 and 129 Infantry Brigades. Hill 112 was thus once again to become a 'no man's land'.[12]

At 2045 hours Headquarters 29 Armoured Brigade noted in their war diary: 'In view of conc of large Pz force on rt flank of br head, warning order reced and issued that the Bde would move back to area NORREY-EN-BESSIN during night.' Before the withdrawal orders could be issued to the motor companies, the Germans attacked again.

Following the extremely heavy air and artillery attack II SS Panzer Corps had suffered during the afternoon, *Oberführer* Harmel, commander of 10th *Frundsberg* SS Panzer Division, gave orders for the transfer of the *schwerpunkt* (focus) of his advance on Hill 112 to the right flank. Further away from observers in the Odon valley and as darkness fell, without direct observation the British artillery would be far less effective, and Allied aircraft would also be absent from the skies. Avoiding Hill 113, the main approach to Hill 112 would be from Évrecy, further south via Avenay and the valley of the Guigne.

As darkness fell, II SS Panzer Corps' attack resumed. South of the Odon, the *Frundsberg*'s 22 SS *Panzergrenadiers* were again probing the Odon bridgehead but further south 21 SS *Panzergrenadiers* holding the *schwerpunkt*, advanced via Avenay. Before long, as dusk settled on Hill 112, 8 RB's adjutant was recording in the war diary at 2115 hours: 'Enemy patrols moving towards our positions from WEST and SOUTH.' As far as those on the hill were concerned, these were not just patrols:

> At last light, some eight tanks, reported as Tigers, though not confirmed as such, together with something over 150 infantry, were observed advancing from the west towards the Company positions, by No. 3 Section. Very lights were going up and machine-gunning broke out. The long-awaited counterattack for which we had waited all day, was coming in at last. Strange as it may seem, there was a certain relieving of tension, for anything was better than that perpetual suspense, not knowing when or whence the attack would come. Now it was coming in, we knew where we stood. Artillery support

was again enlisted, and within a matter of minutes, for time was urgent, a devastating barrage was brought to bear on to the advancing infantry. For nearly a quarter of an hour the boot was on the other foot; the gunners succeeded in wreaking complete havoc on the enemy on the ground. The tanks finding their ground support virtually liquidated, withdrew, together with what remained of the infantry, and, with darkness now complete, we awaited our next call to action.

II SS Panzer Corps, under pressure to make progress, had attacked before the two British artillery observers on Hill 112 were blinded by night. The result was that, with the considerable flexibility of the Royal Artillery, another advance-halting bombardment was brought down on the SS troops.

The threat of attack and the presence of enemy patrols to the west at 2100 hours no doubt disrupted battle procedure for the withdrawal from Hill 112, where the defenders of G Company now stared out into the full darkness:

Meanwhile, 10 and 12 platoons, forward on the slope were still observing to the south. There was a wire fence running across their front, about 100 yards distant, the posts of which were spaced about 10 yards apart. In the darkness, as their watchful eyes gazed down through the wires, the posts seemed to take human form, wore German helmets, and carried rifles, according to more than one observer. 'So full of shapes is fancy.'

A knocked-out Sherman of 29 Armoured Brigade being used by a German observer in a later battle for Hill 112.

It was not until 2200 hours that 8 RB's battalion headquarters received a warning order: 'Bde warns us that we may be ordered to withdraw during the night to NORTH of R. ODON. From then on shelling continues spasmodically. The expected attack on our posn is not launched, though enemy patrols remain active.' When confirmation of the withdrawal came from brigade headquarters, such was the shelling that Colonel Treneer-Michell could not summon an O Group but instead again braved the shellfire to visit each company head-quarters, including H Company in the Orchard:

> At about midnight there was a loud crash of mortar bombs, and the Com-manding Officer hopped into the Company Headquarter trench. Stewed fruit was being eaten at the time, and washed down with whisky from a flask. He refused a hospitable offer to share with us this most precious refresh-ment, but brought the disturbing news that we might have to move back ... An enemy patrol had also come up to the wood, and was driven off by the anti-tank platoon. We would, therefore, have been happier to stay where we were. There was another whirr and crash of 'Minnies' – and the Com-manding Officer popped out of the trench into the night.

It is possible that during the action with the enemy patrol, Lieutenant Ireland-Blackburn, one of the two anti-tank platoon commanders who were forward in the Orchard, was wounded and evacuated, leaving Lieutenant Elgood with the difficult task of commanding both widely dispersed platoons.

For some time, there was clearly some indecision and uncertainty in both 3 RTR and 8 RB as to what brigade headquarters wanted them to do, no doubt exacerbated by the presence of the enemy. The confusion is revealed in the diary of Trooper Thorpe, who was with the veteran 3 RTR on Hill 112: 'Warning

A carrier and 6-pounder anti-tank gun knocked out in the Hill 112 area.

Order received: Abandon tanks after destroying gun. But no action until confirmed. New orders: Retreat, taking the tanks with us. Does anyone know what is going on?' At 0100 hours 8 RB's war diary records: 'Bn receives orders to withdraw covered by 3 RTR,' and that between 0115 and 0230: 'Withdrawal commences in order H, F, G Coys. Bn RV at 954635 [In the valley at the junction of the Roman road and the Fontaine-Étoupefour–Baron road].' Up on Hill 112 conditions were difficult:

> An hour later, the word to go came through. The tanks from the wood moved out first and then would cover us out. Orders to move had again to be given by walking round to each position in turn. As not every position was in the wood, this took some time. It was also one of the causes for an unfortunate misunderstanding with the anti-tank platoon which resulted in three guns being left behind, which subsequently caused a lot of trouble. Another unfortunate thing was that all the Company cookers and some tins of food had to be left behind, as each platoon was to carry only its weapons, in case it should have to fight its way back.

The ramifications of the expectation of having to fight their way out through German patrols and the consequent abandoning of weapons and equipment is covered in the next chapter. Among those left behind was a section of 13 Platoon near the Rectangular Field. Eventually realising they were on their own, they exfiltrated back into the valley.

Having gathered and accounted for the majority of the battalion, at 0300 hours the adjutant noted: 'Message received from Bde that it is vital that the Bn move off at once.' The movement and the sound of tracked vehicles around Esquay and to the south of Hill 112 heralded a renewed attack and General Roberts wanted the final element of 29 Armoured Brigade to be well clear of 159 Brigade's positions by dawn:

> We reached the RV – the burning half-track of the night before – at about 0245 hrs. The battalion column was lined up by Companies, ready to move back over the bridge we had crossed only 57 hours before. Cuff chose this moment to drive his carrier over an iron post, which jammed the track and then broke it. After much trouble with this, the column moved off without the few of us who were with the carrier. We set it on fire and pursued the Company in the two mortar carriers, but soon found ourselves lost in a small farmyard. Cpl. Beaumont then took charge of this party and guided us to the bridge, which we reached as the last vehicles were going across.

The war diary recorded at 0310 hours that the 'Bn moved off and crosses R. ODON by Br at TOURMAUVILLE. Withdrawal completed without further action or enemy interference.' In H Company: 'Everyone was very tired and the drive back was depressing. We passed through many derelict villages and it seemed that we had given so much and lost so many of our friends and done so much destruction – all in vain.' Captain Bell noted: 'Just over an hour before dawn we were again ordered to withdraw. With most of the vehicles already gone, it was once more a case of climbing on to anything that moved.' In a further example of the conduct of the withdrawal. 'About half the Battalion went right back to Cully, where our "A" Echelon was still stationed and which seemed our natural home from home, because no one knew where we were meant to go.'

By 0630 hours, 29 Armoured Brigade, or most of it, was in a concentration area just east of Norrey-en-Bessin.

The regimental historian sums up the First Battle for Hill 112:

Hill 112 was for the 8th Battalion its first experience of war. The operation combined the worst features of the fighting at this time. While the Germans could not maintain their position on the hill, they could bring very heavy fire to bear on anyone who turned them off it. Most of the Riflemen were

Riflemen of 8 RB provide an action scene for Sergeant Laing of the Army Film and Photographic Unit on 29 June 1944.

subjected to shelling and mortaring without ever seeing a German, and our tanks were picked off by an enemy whose better armament and skill in concealment gave him an advantage over our troops who were advancing across the open. The thick, wooded, leafy bocage country handicapped our Gunner observation posts as well as the armoured regiments; and as vehicles burnt and bombs and shells exploded in the cramped area of Hill 112 the air became loaded with a fine, grey powder, covering men and vehicles and the bodies of cows so that the wounded looked even more ghastly than usual, lying among the smouldering shell-holes and splintered trunks of trees. Casualties in this first battle had been heavy, amounting to three officers and thirty-four non-commissioned officers and riflemen in 'G' Company alone. Already two company commanders had been wounded.

No matter how awful the conditions on Hill 112 were, there were those who rose above it all. Lance Corporal Piggot earned the Military Medal on Hill 112. His citation reads:

L/Cpl Piggot has fought with 'F' Company, 8th Battalion, from Normandy from June 1944, to the Baltic, May 1945. As Rfn Piggot, on 28th June 1944, he took part in the attack on Hill 112 and was a constant source of inspiration to the rest of the men in the platoon. Very few of them had been in action before, and Rfn Piggot in his first battle showed himself to be a good leader in action in addition to being a very brave man.

Luftwaffe anti-aircraft gunners who had been taken prisoner are photographed by an AFPU south of the Odon, 28 June 1944.

Chapter Five

Between Battles

> Our time at Norrey was spent licking our wounds from Hill 112 and trying to take stock of our freshly gained experience. I don't think any troops could have had a sterner initiation into the un-pleasantness of battle than we had on that legendary hill. [Captain Bell]

Captain Bell arrived in a field to the west of Norrey-en-Bessin on the morning of 30 June:

> This was to be our laager area while we reorganized. The Company transport was formed into a single column, and we dug in beside our own vehicles. The scene around us was not uplifting, for it was here, soon after D-Day that the Canadians had fought a tough engagement in the initial and abortive break-out attempt. Several knocked out Mark IVs lay about, suspension-deep in the cornfields. Here and there, half hidden by the tall, ripening cornstalks, lay a dead soldier, whose blackened face gazed un-seeingly towards the sky. Some were in khaki, others in field grey.[1]

In just four days, during its first action, 8 RB had thirty-six men killed and 123 wounded, approximately 20 per cent of its strength. In the aftermath of EPSOM, a war diary entry covering the period 1–8 July summarises the battalion's activities:

> During this time the Bn was virtually left alone to reorganize. Armd Regts were positioned in areas from which they could counter-attack any threat to the flanks of the ODON bridgehead, and Motor Coys were out with them for a few days at a time, but the Coys were not engaged and we were able to profit from the fact that we had most of the time to ourselves and on the 6th the residue [of twenty-four] men and vehs arrived. We were still [ninety-five] down in men but reorganised and battleworthy Coys reverted to command. The Bn as a whole took up a defensive posn to protect the West flank of the ODON bridgehead, with a counter-attack role in the area.

From this point in the campaign, until motor battalions in the UK started being broken up, the number of replacements received failed to keep pace with the casualty rate.

Sergeant Hicks had been left behind in Aldershot when the battalion left on 8 June and with the other First Reinforcement of 8 RB had joined 34 Reinforcement Holding Unit of 101 Reinforcement Group. They had been landed during

Riflemen of F Company 8 RB photographed in leaguer below Hill 112 during Operation EPSOM.

EPSOM and on 30 June they began the process of moving down the line of communication to join the battalion:[2]

> The landing craft eventually came out and we boarded them from the gangways slung over the side of the ship and within a quarter of an hour our craft grounded on to the beach, the whole of the flat bow was lowered to form a sloping ramp and all we had to do was run off up the sandy beach without so much as getting our feet wet.
>
> Evidence of the D-Day onslaught was to be seen all around the beachhead, burnt-out vehicles, smashed guns and bits and pieces of all kinds of equipment littered the scene. Blasted trees and buildings, torn barbed wire defences, temporary graves with rifles stuck into the ground by the muzzles and with a German or British steel helmet stuck on to the butt end to crudely denote the nationality of its occupant.

Immediate dispersal from the beach was essential so we were assembled and marched off to the outskirts of Bayeux to a given map reference which turned out to be a very large field in which had been dug numerous slit trenches by our predecessors. We stayed there for a few hours resting and eating part of our emergency field rations before moving on to occupy some derelict farm buildings. Considerable gunfire could be heard in the distance during the couple of days we spent there and although none of it came in our direction we were mindful of the fact that this was for real and not a mock battle somewhere in England that we had experienced so often in previous years.

Orders were received to move on to our last RV area some 2 miles further inland, where we waited for transport from our battalion to pick us up. We had had no firm news of the battalion but a rumour had reached us that it had already been in action.

Transport finally reached us on 30th June with Captain 'Tanky' Townsend [OC Headquarter Company and B Echelon's commander] in charge and our questions were soon confirmed by him that 8 RB had been engaged in an abortive two-day battle to capture Hill 112 ... So, 1st July found us reunited with the old company in an orchard near Putot-en-Bessin. It was a fine hot summer's day and early in the afternoon when we arrived. Everybody was relaxing in the sun, either writing letters, listening to radios, cleaning and sorting out weapons and this, that and the other and we certainly received a very warm welcome ...[3]

Sergeant Hicks' experience of returning was typical of the early replacements, as he was posted to his old company, H Company, as sergeant in the 3-inch mortar detachment. For those who joined as reinforcements later, they knew nothing of the battalion, were often from a different regiment and were increasingly not motor battalion trained. Despite reinforcement, 8 RB started its next battle no fewer than 103 riflemen under its establishment.

As was the case across the Second Army, there were departures following most formations' first battles. Some, such as Brigadier Sandle of 159 Brigade, proved to be temperamentally unsuitable for battle, while others did not perform as well as expected. The 8 RB war diary entry for 8 July simply noted: 'Lt. Col. E.D. Treneer-Michell left us and his place was taken by Lt. Col JA Hunter MBE MC of the 60th [King's Royal Rifle Corps (KRRC)].' General Roberts noted as a result of the confusion during the withdrawal from Hill 112 that 'some aspects of the performance by 8th Rifle Brigade were not entirely satisfactory and some equipment was lost so that [Brigadier] Roscoe Harvey felt he had lost confidence in the CO and recommended that he should be replaced'. Writing in a letter of congratulations to General Roberts, the corps commander, by singling out the armoured regiments for praise, intimated that he concurred:

I consider the operations of 29 Armoured Brigade east of the ODON were highly creditable and showed the highest standard of training attained by all units. Not the least creditable part was where the arrangements for the night withdrawal which were carried out without a hitch as far as the armoured regiments were concerned.

All three senior officers in 8 RB's chain of command were from an armoured background and it could be argued that back in their command posts they did not fully appreciate the seriousness of the infantry situation in the Orchard. Trooper Thorpe's comment during the withdrawal from Hill 112: 'Does anyone know what is going on?' surely reveals a state of more general confusion emanating from 29 Armoured Brigade. A 'loss of confidence' in battle is, however, a loss of confidence. General Roberts concluded that: 'It just goes to show that exercises, even as realistic as Exercise Eagle had been, cannot really test men for battle.'[4]

The newly arrived 28-year-old Lieutenant Colonel Hunter had looks to match his age, as Lieutenant Stileman recalled:

I saw this youth advancing towards me. Assuming that he was a replacement subaltern, after our EPSOM casualties, I was about to address him in familiar subaltern terms, when I noticed, at the last minute, that the top 'pip' on his shoulder was in fact a crown and that he already wore the ribbons of

VIII Corps' commander, General Sir Richard O'Connor.

MBE and MC won in the desert campaign. I was therefore dutifully polite to my new commanding officer!'

At the same time as Colonel Treneer-Michell's replacement, Major Dickenson, also of the KRRC, came to command H Company and Captain Bell took over G Company.

The Counter-Attack Role

At the end of the first week of July, the attacks of II SS Panzer Corps were over and the Germans were concentrating on relieving their panzer division in the line with four newly arrived infantry divisions. The Second British Army now switched its main effort further east with operations CHARNWOOD, the capture of Caen and JUPITER, a second attempt to secure Hill 112 and the crossings of the Orne. Meanwhile, to the west, VIII and XXX Corps largely stood on the defensive. However, with II SS Panzer Corps still active on the western flank of the EPSOM Salient, 29 Armoured Brigade was responsible for providing reserves to back up the infantry brigades and 4 Armoured Brigade deployed in defensive positions. The 29th's armoured regiments and their motor companies were deployed in turn but were not committed to battle. They endured regular shell and mortar fire and suffered casualties as a result.

On 7 July, 29 Armoured Brigade received a warning order: '8 Corps is holding and strengthening its present posn with a view to defeating any counter-attack that the enemy may launch against the corps' front.' To this end the brigade, less the F&F Yeo, which was already forward with 160 Brigade, was to relieve 4 Armoured Brigade in the counter-attack-role in support of the 53rd Welsh Division. The motor companies reverted to command of 8 RB and the battalion deployed forward as a whole to take over positions north of the long straight Fontenay road the following day. The brigade plan was that 23 Hussars were to be prepared to counter-attack onto the Rauray spur, while 3 RTR and 8 RB would form two company/squadron groups to recapture Cheux and Le Haute du Bosque in 71 Brigade's area if lost to the enemy.

Battle procedure began at once, with the CO visiting fellow motor battalion 2 KRRC, which 8 RB were to relieve, the following morning. Once the relief was complete, Brigadier Harvey gave his orders for the counter-attack role at midday to the COs of the RTR and 8 RB. In addition to the COs, the company, squadron and battery commanders were present, along with the second in command of both G and H companies. Lieutenant Colonel Hunter was holding his first O Group that evening when he was interrupted by 'a distant throb of aero engines that filled the sky.' Captain Bell recalled that:

> Looking up, we saw approaching a great black cloud of RAF 'heavies' on their way to bomb Caen, prior to the [British and] Canadians launching their attack on this old Normandy town, so vital to us for the success of future operations. Order group was automatically suspended while we watched the amazing spectacle of the seemingly endless stream of bombers go in to drop

their loads. Anti-aircraft fire, which came up quite strongly at first, was soon smothered and great fires began lighting the sky. Dusk was approaching as the last bomber turned and started off for home, and we envied the crews who would soon be sitting down to a good meal in England.

The following day, with the luxury of time, as noted in 3 RTR's war diary: 'Rehearsal with Sec comds 8 RB and Tk Comds 3 R Tks held in CHEUX area. Rehearsal of tasks as laid down by Comd 29 Armd Bde for counter-attack role with 8 RB in support of 71 Inf Bde.'[5] The rehearsals were, however, not universally popular. Captain Bell continued: 'We rehearsed our counter-attack role, which appeared rather remote and caused a certain amount of umbrage to be taken by the forward troops, as movements in their area brought down a hail of shells and mortar bombs.'

The days were spent with the company's trucks parked up along hedge lines in the shallow Mue valley, with the riflemen, thanks to the regular mortaring,

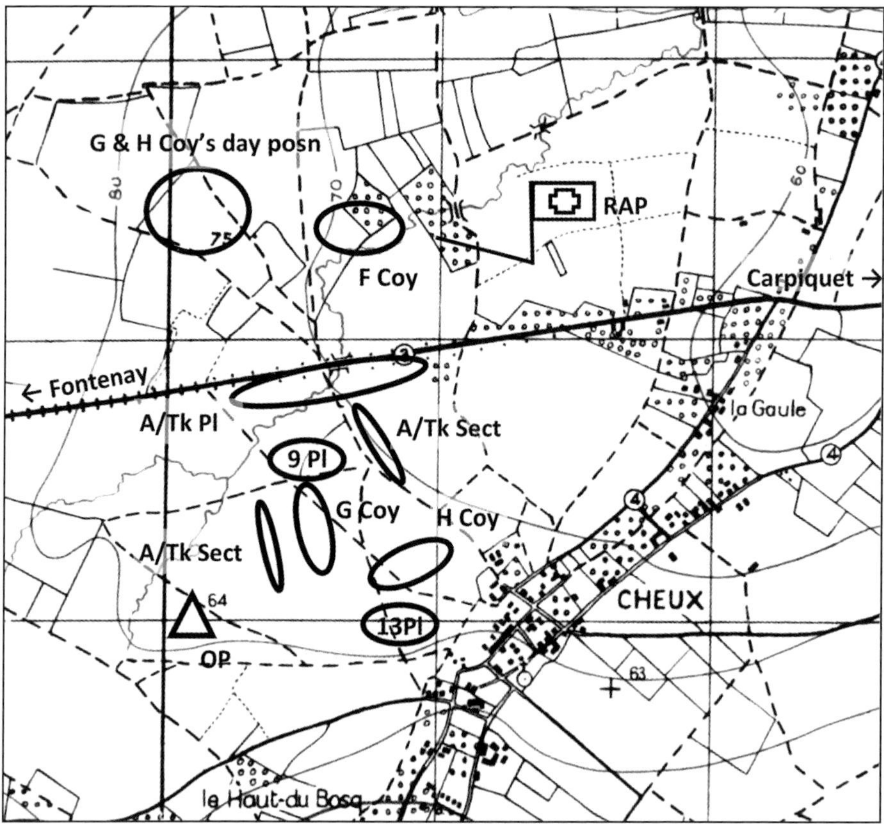

The deployment of 8 RB and 3 RTR in the counter-attack role, 8–14 July 1944. Taken from a trace in the battalion's war diary. Note that the myriad of Royal Artillery gun positions that surrounded the battalion are not shown.

dozing in the trenches they had taken over from the KRRC. F Company recorded that: 'Apart from occasional shelling, which put a hurried stop to the birthday party of one of our officers, and our first attack by a flight of Messerschmitts in a ground-strafing role, there were no incidents' but also on the theme of aircraft, 'Some over-enthusiastic gunners of the Light Anti-Aircraft brewed up their own Petrol Lorry in mistake for a Messerschmitt.'

At last light, however, G and H companies went forward to occupy a tight defensive position on the ridge just north-west of Cheux. This position, which was too exposed for occupation by day, was designed to add depth to the defences and to protect the RTR's tank harbours. E Company remained in reserve in its orchard throughout.

During the first week of July, XII Corps had arrived as part of the final stage of the build-up of Second British Army. Consequently, real estate in the tighter than anticipated lodgement was becoming a real issue and 29 Armoured Brigade found itself in an 'artillery park' of considerable size. According to H Company: 'It was here that we were able to appreciate the strength of 2nd Army's artillery, for we were positioned in the middle of what seemed to be "all the guns in the world".' As the heavy and medium guns of 3 AGRA arrived, the noise in the artillery park grew:

> Night and day this inferno went on, swelling to a mighty crescendo at the end of our stay, when the 15th Scottish Division went through us to battle beyond [Operation GREENLINE]. The noise that night was terrific, but despite the discomfort of having to listen to it, it was most satisfactory to know that the Germans were at the sharp end of it and were suffering far greater discomfort than we were.

The days spent in the counter-attack role passed without the plans being tested by the enemy.

During the days back in the battalion area there were occasional trips to ad hoc bath units. After a week without clothes off his back, Sergeant Hicks described one such trip in a letter home in early July:

> Actually, on Tuesday we went back for a bath but you would laugh if you saw the Heath Robinson contraption which served as a shower bath. They had suspended three old oil drums on each side of a truck, each pivoted near the top and half a dozen holes punched near the rim to act as a rose. So all they had to do was pour a bucket of hot water, which was our allowance, into the drum and tip it up as we wanted it – and oh boy was it good.

Unlike visits to proper RAOC bath and laundry units, there were no changes of underwear, etc.

The Situation in Mid-July

During 16 June the battalion briefly returned to a field near its old home at Cully, but despite security it was soon apparent that 'something was up' when the

Probably 8 RB's noisiest neighbour, a 7.2-inch howitzer of a heavy regiment being prepared for action in a Norman orchard.

Commanding Officer was summoned to receive orders for what was to be General Montgomery's next 'colossal crack', Operation GOODWOOD. Montgomery's campaign plan was nearing its climax, the breakout by the First US Army in the west. To ensure that General Bradley's divisions attacked in the most favourable circumstances, German armour needed to be forced to confront the British and once on the Second Army's front, kept there. The British attack was to be delivered on 18 July, but sequencing with the Americans was not good, thanks to the bocage delaying General Bradley, the weather and mis-bombing, all of which paused Operation COBRA until 25 July.

With three armoured divisions in hand and the intercepted German signals saying that, having suffered 100,000 casualties since D Day, they were 'nearing breaking point', now was the time to produce a threat south-east to the Seine and Paris to which the enemy would be forced to respond. Of particular concern was the fact the 272nd Infantry Division had arrived with I SS Panzer Corps and were relieving the panzer divisions. The *Hitlerjugend* had been largely extracted from the line for the first time since their committal to battle on 7 June and the *Leibstandarte* were also starting to be relieved by the Wehrmacht infantry. From Montgomery's perspective, these divisions could not be allowed to be reconstituted as a German operational reserve and needed to be brought back and fixed against the Second Army. In addition, if significant ground could be made to the south of Caen, this would go some way to answering the growing criticism of a lack of progress in capturing airfield sites. Another factor in Montgomery employing his armour as GOODWOOD's centrepiece was that following earlier actions, the 7th and 11th armoured divisions had been almost fully reconstituted, and they had been joined by the Guards Armoured Division, together fielding some 877 tanks. In contrast, the infantry divisions, as a result of fighting south of

The Sherman was the standard tank of both the 11th and Guards armoured divisions.

Bayeux, EPSOM, the battle for Caen and the 43rd Wessex Division's renewed attack on Hill 112, had all suffered much higher than forecast casualties.[6] Both the reinforcement holding units in Normandy and training units in the UK were emptying rapidly; a month into the campaign, a manning crisis for Second Army was already looming.

German Defences East of the Orne

The Germans had always expected that the British would attempt to break out from their lodgement east of the Orne. Their analysis of what was presented to them by Operation FORTITUDE was that the landing by the fictitious FUSAG would take pace once 21st Army Group reached the Seine. On the morning of 17 July in Army Group B's situation report for OKW covering the previous week, Rommel's Chief of Staff, General Dr Speidel wrote:

> The British 2nd Army has not achieved its aim of breaking through and operating in the open areas after the occupation of Caen. The attack was not followed up owing to heavy losses. However, regrouping and preparations on an increasing scale were observed towards the end of the week. The well-known operational intentions of the Montgomery Army Group still appear to exist. The British 2nd Army is clearly concentrated in the area of Caen and to the South-West and will carry out the thrust across the Orne towards Paris. The local attacks on July 15th between Maltot and Vendes [Operation GREENLINE] may be the prelude to the large-scale attack which is expected from the evening of the 17th for making a breakthrough across the Orne.

Consequently, the Germans had had plenty of time to prepare a defence in depth south of Caen. The potential Allied axis of attack south from the Orne bridgehead to the high ground beyond cut across boundaries of two German corps, the LXXXVI to the east of Caen and I SS Panzer Corps to the south of the city.

The first of the German defensive layers was almost sacrificial, being provided by the 16th Luftwaffe *Felddivision*.[7] This was one of the first reinforcement formations to arrive in Normandy, replacing elements of 21st Panzer Division astride the Orne. Never a very strong division, it had been further weakened in defending Caen during Operation CHARNWOOD, when one of its infantry regiments suffered 75 per cent casualties. What was left of the division was now deployed across the southern front of the British lodgement east of the Orne, from the river in the west to its junction with the 346th Division at the Bois de Bavent. However, with as few at 1,500 infantry, it was indeed little more than an outpost line.

The second line was the main defensive position held by *Kampfgruppe* von Luck, based on *Panzergrenadier* Regiment 125, which had under command the remaining Mark IVs of 22 Panzer Regiment and the Tiger Is and IIs of 503 *Schwere* Panzer Battalion. However, the key unit in this line was *Sturmgeschütz* Battalion 200, which occupied a series of mutually supporting positions,

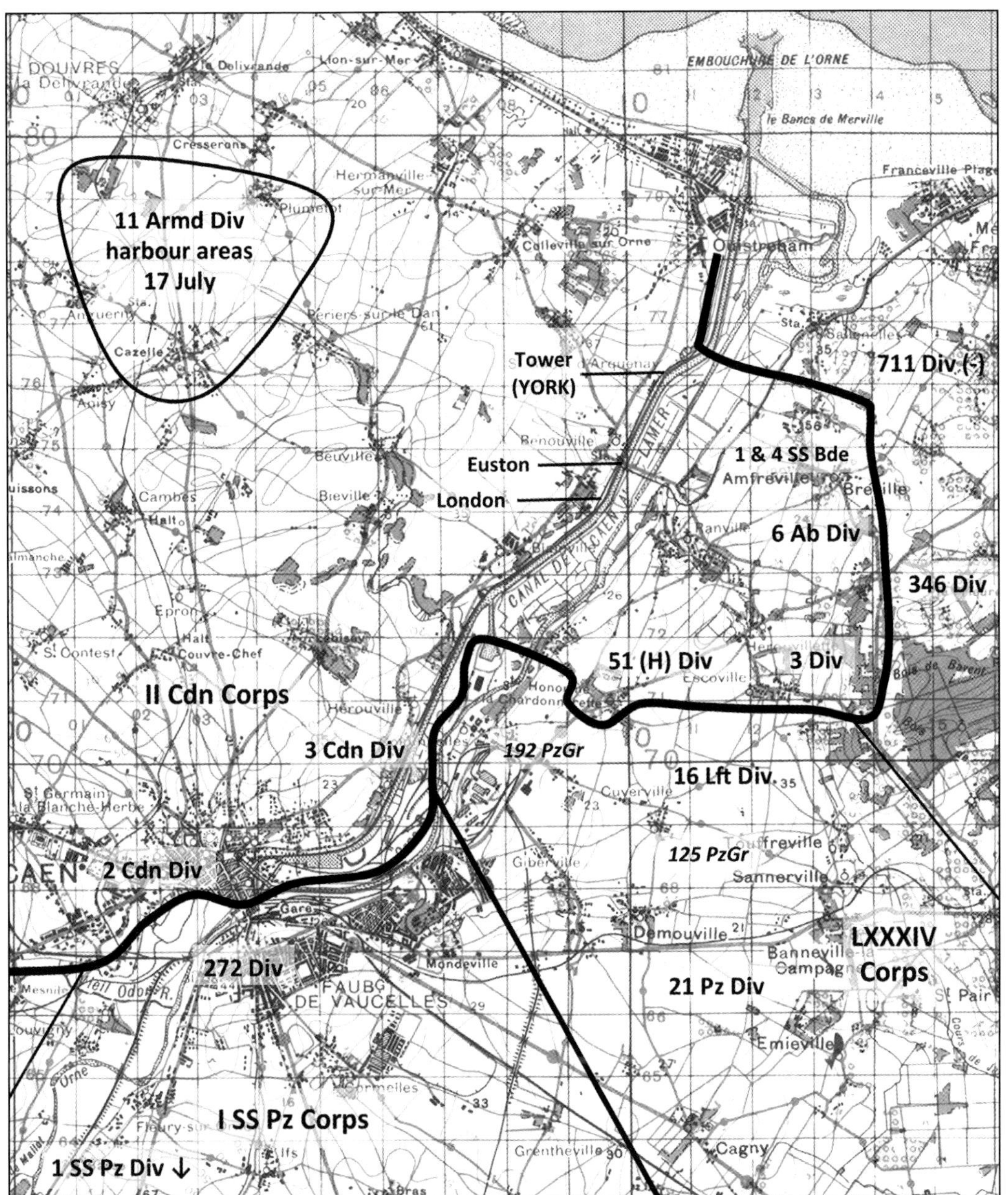

British, Canadian and German formations around Caen and the Orne Bridgehead on 17 July 1944.

in the cover of villages and folds in the ground, spread across the corridor down which any breakout from the British position was bound to take. Their aim was to halt an attack in this main battle area before it reached the Caen–Vimont railway 3 miles behind the German forward defence line.

Some 6 miles from the Allied start line, the next position was on the Bourguébus Ridge, where the 21st Panzer Division's reconnaissance and pioneer battalions were deployed to protect the corps artillery, which consisted of a modest number of guns but a prodigious array of *nebelwerfers*. The LXXXVI Corps alone had 194 artillery pieces, *272 nebelwerfers* and seventy-eight anti-aircraft and anti-tank guns. South of the Bourguébus Ridge elements of the 1st *Leibstandarte* SS Panzer Division were I SS Panzer Corps' reserve.

The GOODWOOD Plan

The essence of the British plan was to speed a substantial armoured force through the German defences and into the rear area with the aid of a massive aerial and artillery bombardment. To that end, with many fewer hedges, the ground south of the Orne Bridgehead was considered to be 'good tank country' of mainly open gently undulating arable fields, where both speed and sweeping manoeuvre

---

Sturmgeschütz Battalion 200

Major Alfred Becker was an industrialist, mechanical engineer and reserve artillery officer who began exploiting captured and knocked-out enemy equipment during the 1940 campaign, when he motorised his battery with captured Dutch vehicles. Later the same year, while on coastal defence he took the chassis of abandoned British Mk VI Light Tanks and converted them to mount his 105mm leFH 18 guns, producing a self-propelled battery, which proved successful. British Bren gun carriers were other early conversions.

With the demands of the Eastern Front, the German firm of Alkett had started to convert Renault R35 tanks to 47mm *Sturmgeschütz*, but by 1942 they had little effect on Soviet tanks. As the leading exponent of conversion of captured armour, Major Becker was summoned back to Germany to find a way of mounting the 75mm PaK 40 on French armour. He demonstrated the Marder 1 to Hitler and his staff, who saw the advantages at a time of increasing shortages. Consequently, when the 'New' 21st Panzer Division was raised in May 1943, as the manpower and equipment crisis of that year bit, OKW instructed that:

> Equipment (aside from communication kit) and vehicles are not allotted. It is expressly forbidden to requisition material for 21. Panzer Division in any way. The necessary equipment and motor vehicles are to be secured exclusively from French booty from OB-West.

Becker eventually produced a total of 1,800 AFVs in the *Baukommando* Becker (Construction Staff)[8] works near Paris. Of these, some went to *Schnelle* Brigade West[9] and then to 21st Panzer Division when the brigade was increased to a full division in August 1943. The re-raised division was commanded by the newly promoted *Generalmajor* Feuchtinger.[10] The panzers were initially all French Somua vehicles, but these had been virtually all replaced by Panzer IVs by June 1944. The rest of the division's vehicles, some 447 of them, were, however, re-engineered 1940 vintage French armour, including those that equipped *Sturmgeschütz* Battalion 200.

As an artilleryman, and having worked to produce self-propelled anti-tank guns, Becker proposed an experimental unit that fused both types of weapon. His old commanding officer from 1940, *Generalmajor* Feuchtinger, readily agreed to five batteries of six 75mm anti-tank guns and four 105mm guns, based on the Renault FCM or the Hotchkiss H39. In addition, there were sundry French-manufactured command, OP and ammunition-carrying vehicles.

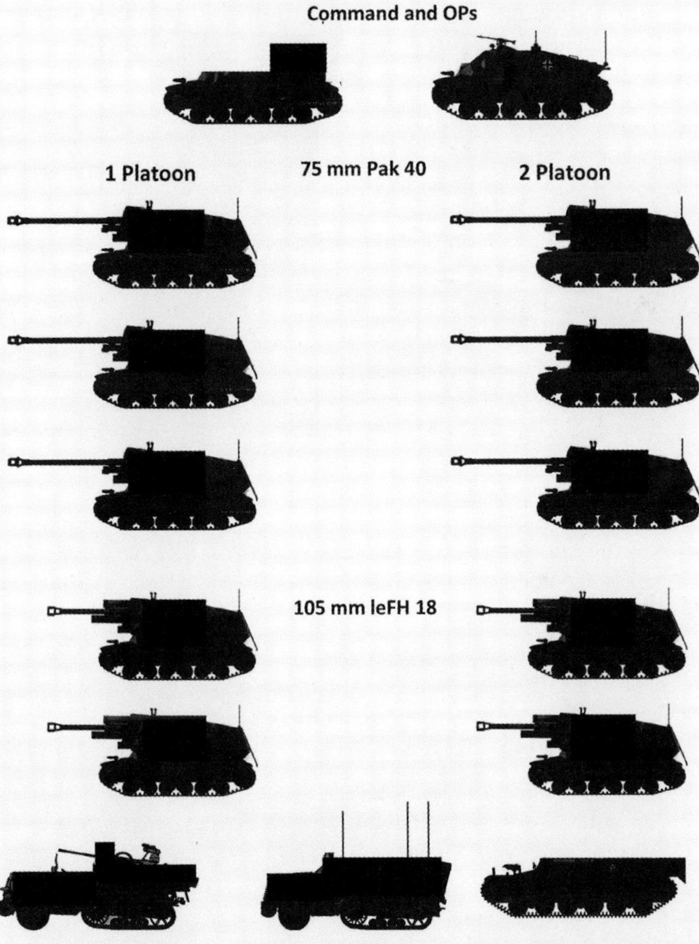

The Battalion's fifth battery was incomplete, at the beginning of the campaign having four leFH 18 but only two PaK 40s.
Other batteries were not at full strength on 17 July due to combat losses since 6 June and mechanical failure.

75mm PaK 40 auf *Geschutzpanzer* 39 Hotchkiss (f).

10.5cm leFH 18 (Sf.) auf *Geschutzwagen* 39 Hotchkiss (f).

10.5cm self-propelled guns with command and OP vehicles in the foreground and supply carriers behind.

were practical propositions. However, with a strongly emplaced and skilful enemy, such ground with its long and open fields of fire, is also invariably 'good anti-tank country.' As VIII Corps' GOODWOOD monograph pointed out, there was plenty of cover for anti-tank gunners in the paddocks and gardens that surrounded the villages and farms. From these, at ranges of between 1,000 and 2,000 yards the guns had interlocking and overlapping arcs of fire covering the ground over which the corps was to advance.

To deal with the German defences, which were known to be based on strong-points deployed in some depth, a preliminary bomber strike was requested. The decision to use the heavy bombers of RAF Bomber Command and the US Eighth Airforce is described in VIII Corps' GOODWOOD Monograph:

> Owing to the large number of strong points which dominated the area of the initial advance, it was considered that the operation would not be feasible unless supported by maximum air support to augment the arty fire power. A feature of this air support was the election of the types of bombs to be used for different objectives.[11]

In the centre of the attack was 'Tank Alley', down which the armour was to advance before fanning out. In this and other areas where armoured manoeuvre was planned, bombs with instantaneous fuses that avoided cratering were to be used. To the flanks, in areas through which dismounted infantry were to attack,

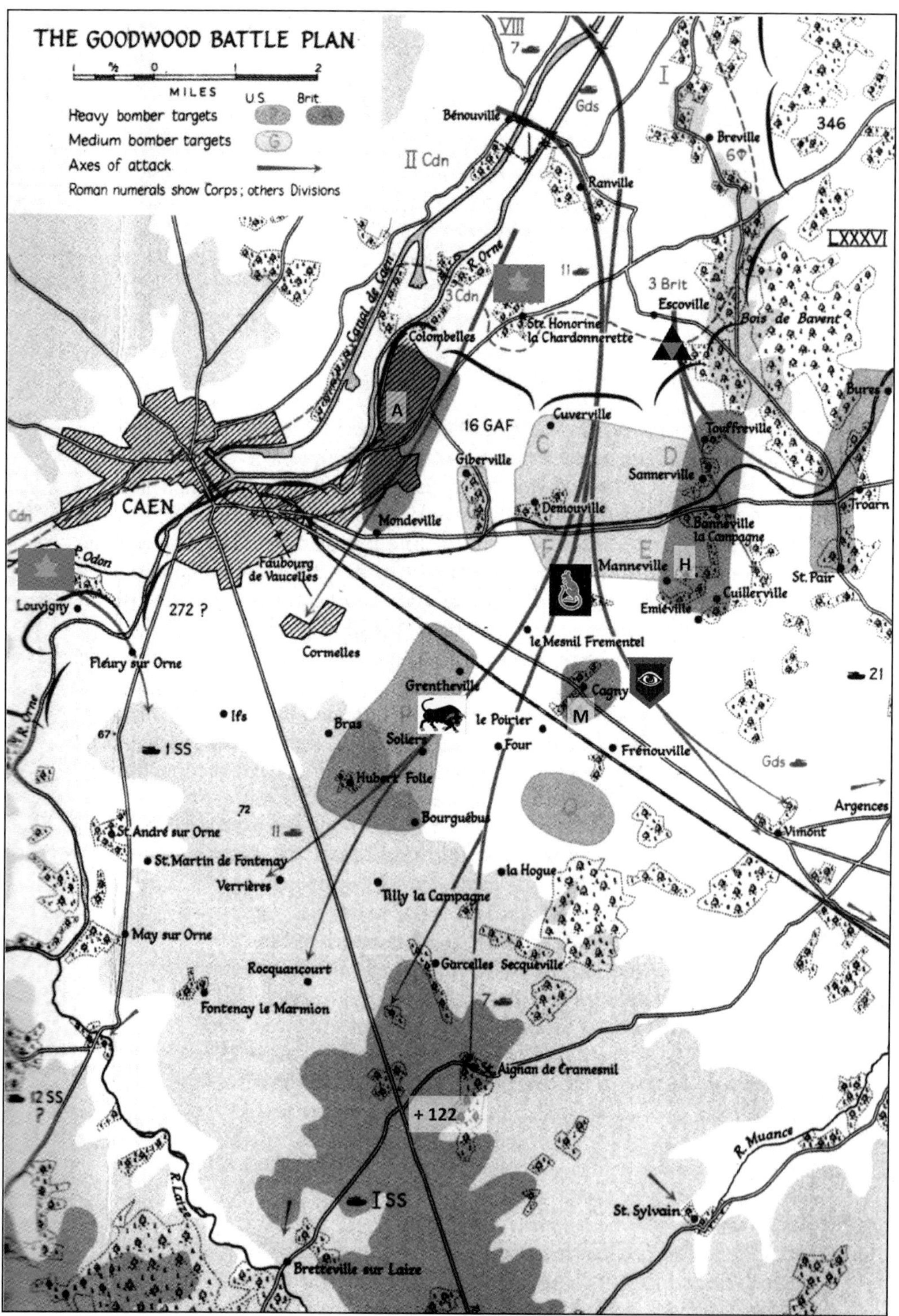

The official history's map of the GOODWOOD plan. Not shown on the map are strikes by 83 Group RAF on roads in order to isolate the battle area.

the bombing was maximised for destruction and, therefore, cratering. The air strike was to begin at 0545 and last for almost two hours. Once the bombers departed, a barrage would be fired by the massed artillery of three AGRAs and the 25-pounders of seven infantry divisions. Much of the latter was to provide a creeping barrage behind which 8 RB and the tanks would advance at 5mph.

With just 2 miles between the Bois de Bavent and the Colombelles Steel Works, and the restricted area in the Orne Bridgehead, VIII Corps' three armoured divisions would attack in sequence. The advance would be led by the 11th Armoured Division, which had the experience of EPSOM to build on, and once out of 'Tank Alley' they would swing to the right. Following them, the newly arrived Guards Armoured (GAD) was to come up alongside the 11th and swing left, while bringing up the rear was the 7th Armoured, the most experienced division. From the centre they were expected to assist where there was difficulty and/or exploit south to Point 122 on the Falaise Road. The divisional history of the 11th Armoured noted: 'Whether or not this advance would be pressed to Falaise depended on the enemy's reaction ...'

General Roberts was given a plan by General O'Connor with little scope for change. On his right, the infantry battalions of 159 Brigade were to attack the villages of Cuverville and Démouville in turn, while with just 1,500 yards between these villages and the 3rd Division's boundary to the left, 29 Armoured Brigade's advance would be heavily constrained.

Preparation for Battle

After only a few hours at Cully, the motor companies rejoined the armoured regiments, as a part of a regrouping that involved not only 8 RB's companies but guns, flails and AVREs of 79th Armoured Division. Moving 11th Armoured Division group alone involved 3,500 vehicles and was a major undertaking in both planning and execution. After midnight on 16/17 July, 29 Armoured Brigade started out on the first stage of its move on routes CALF, CAT and RAT, east to harbour areas several miles short of the canal and river. For the headquarter element of 8 RB and its attachments, the first phase of the movement plan was a 13-mile drive to a harbour area between the villages of Cressons and Plumetot. Night moves were exhausting for drivers and vehicle commanders but not exactly restful for the riflemen in the back, who could only doze uncomfortably. Crossing the rear of I and the II Canadian Corps, it took a stop-start two and a half hours to cover the relatively short distance to the harbour area.

The F&F Yeo, with F Company and I Battery at the rear of their column, recorded that: 'It was a bad night for moving – dust, mist and a very dark night, all contributed to the difficulties.' G Company's account adds more detail:

It was close on midnight when we set off, nose to tail, in almost pitch darkness. Most of our movement was along tracks, and the fact that the route was not clearly marked, coupled with the darkness and the great volumes of dust thrown up by the unending columns of armoured fighting vehicles, made the going very difficult. Kenneth Chabot had gone on ahead as a harbour party, and it was with very great relief that we at last met him waiting for us early next morning.

Lieutenant Chabot and a junior NCO from each platoon went ahead with the rest of the 3 RTR battlegroup's harbour party. Before last light they were briefed on where the company's vehicles were to be parked and they then waited to meet the company at the entrance to the harbour and from there guided the half-tracks into their positions.

On arrival with RTR in a field near Plumetot, H Company recorded that they 'spent the early hours turning the company into a hedge by means of camouflage nets and branches':

> Camouflage Officers were prevalent in the area complaining of our laxness, and it was on this occasion that Rifleman Cuff N. made the rather disrespectful remark 'I am the camouflage man, I carry a roll of scrim etc.', in the pretence of one of these officers. It was while we were pretending to be a hedge that we received our orders for operation 'Goodwood', and they sounded very impressive – 700 tanks, 3,000 aeroplanes, 3 AGRAs (besides our own Divisional Artillery) against the enemy's 200 to 300 tanks. And of course, we also received a NAAFI ration, including our first bottle of English beer.

It had only been days before that, that 8 RB had received its first half slice of white bread from field bakeries now operating in the maintenance area. The arrival of luxuries such as these were memorable points in the campaign. In the case of bread, hitherto they had biscuits AB (Alternative to Bread) white, commonly known as 'hard tack'.

The order for all on the morning of 17 July was for minimum movement, stay under cover and get as much rest as possible. While battle procedure was under way most of the riflemen slept but as Captain, now Major, Bell recalled:

> The Company was roused at four o'clock, and I gave out orders to the order group and then briefed the Company as a whole. All were greatly impressed by the details I was able to disclose, and we felt sure that we would very soon be making headlines for the papers.
>
> We next set to work camouflaging our vehicles. Very nearly the whole terrain across which we were going to advance consisted of cornfields, and so we tied on bundles of corn all over the vehicles, and this turned out later to be most effective. The only danger of it was that an incendiary or tracer bullet might set it alight and cause the vehicle to catch fire, but, weighing everything up, this seemed to be a risk worth taking.

The entry in 29 Armoured Brigade's war diary on the evening of 17 July reads: 'By 2100 hours it was almost certain that the op, which because of the air effort

To ensure operational security, camouflage officers were not happy with simply parking half-tracks and tanks in a hedge or under a tree.

involved, depended so much on the weather, would be on.' Two hours later the brigade was on the move.

To assemble the 11th Armoured Division and to ensure that the rest of VIII Corps could cross the canal and river as rapidly as possible after them, additional bridges had been built. The capacity of the vital Pont de Benouville (Pegasus Bridge) and Pont Tourant (Horsa Bridge), both seized on D Day, had already been enhanced by the building of the LONDON Bridges soon after the invasion, but but the speedy crossing by three armoured divisions required additional Class 40 Bailey bridges to be built. A mile to the north, TOWER Bridges were put in place and alongside Pont de Benouville, EUSTON Bridges were opened by the Royal Engineers. Not only were bridges required but additional roads to serve them. By now it was standard practice for specific tracks to

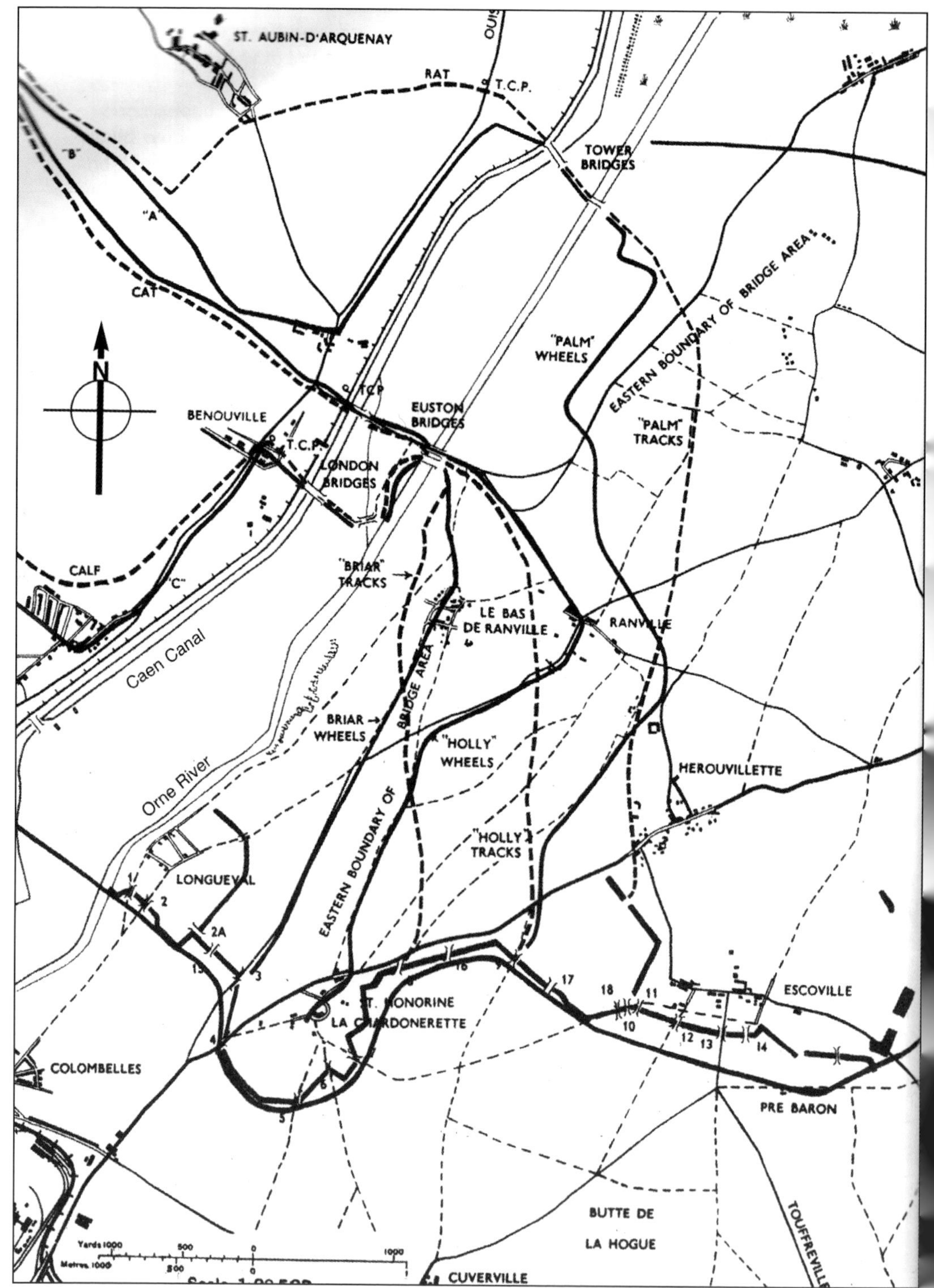

The bridges and roads built to facilitate Operation GOODWOOD. Of the routes west of the canal, CALF, CAT and RAT were for tracked vehicles and A, B and C for wheels.

be made or improved for armoured vehicles in order to keep them off the country roads, which were not built for heavy military traffic, particularly tracked vehicles that would rip them up. In this case, such was the volume of traffic that additional wheeled routes were made as well.[12]

For the final approach, HQ 8 RB, with its half-tracks being classified as wheeled vehicles for the move, followed route RAT to the TOWER Bridges[13] and on to route PALM. The speed was to be 10mph and the distance between vehicles was set as the ability to see the small convoy light under the rear of the vehicle in front. The motor companies crossed with the armoured regiments by the other bridges, with for example G Company crossing the LONDON Bridges in company with 3 RTR. Between 0230 and 0400 hours the brigade, less some stragglers, was complete in the fields north of Ranville, which had on D Day been DZ/LZ November and was still littered with the gliders of the 6th Airborne Division. Major Bell remembered that:

> This was an even more unpleasant drive than the previous night, as in addition to the dark and dust, we this time had shelling added to the menu. However, no one was hit and we eventually arrived at our destination just before dawn broke. Various vehicles went the wrong way, but they were mostly retrieved and the Company was almost complete, but very tired, by the time H Hour came. Kenneth Chabot was still sorting everyone out when I went with Colonel Silvertop [CO 3 RTR] in a scout car to reconnoitre the way to the start line.

The LONDON Bridges were used by 3 RTR and G Company to cross into the bridgehead in the early hours of 18 July.

The trace from the battalion's war diary showing the deployment of the Battalion HQ group.

The Orne Bridgehead was already cramped, with the two special service brigades and the depleted 6th Airborne Division holding the eastern flank and the 51st Highland Division holding the southern front. Add the assault troops of the 3rd Division, plus the vehicles of 11th Armoured Division, and it was extremely crowded. Here the headquarters group of 8 RB and the motor companies waited for H Hour, using the time to get their burners out for a brew and breakfast. Just before 0545 hours the drone of aircraft engines was heard as the first wave of bombers approached.

A Sherman Crab flail tank of 22 Dragoons. Following the break-in battle, the squadron came under command of 8 RB.

Chapter Six

Operation GOODWOOD

While the rifle platoons rested and prepared their breakfast, 8 RB's senior commanders joined officers of the armoured regiments in last-minute recces. Major Bell recorded that: 'I went with Colonel Silvertop (3 RTR) in a scout car to reconnoitre the way to the start line. Just to make things a little more complicated, the 51st Highland Division had laid a minefield, through which gaps had been made for us to get through.' He went on with Major Close of A Squadron, who described how the two of them:

> walked down to the minefield to recce the taped lanes which we hoped would be well signed. There seemed to be no problem – the entrances to the minefield were guarded by immaculately dressed Provost NCOs, all white belts and gaiters. I didn't think we would have any difficulties in getting through the minefield but was a little apprehensive about what we would look like at first light lining up in full view of the enemy [beyond the minefield].[1]

The situation with the minefield was not as good as it looked. Over the previous few nights, the sappers had cleared fifteen lanes through the obstacle but the previous day General Roberts had requested three additional breaches to be made on his front (16–18) in order to speed the deployment out onto the start lines. These additional lanes had been hastily created through standing crops by the 3rd Division's Royal Engineers in the short hours of darkness during the night of 17–18 July. Because the mines had been hastily laid in contact with the enemy, they had not been either laid or recorded accurately and had in some cases been moved or buried by shellfire. The inevitable result was vehicle casualties (see map on page 108 for the numbered minefield lanes).

The two-hour-long bomber strike was universally acclaimed as 'spectacular and morale raising'. To the flanks, 500lb high-explosive bombs were dropped, causing cratering and destruction, while in the 'Alley' through which the armour was to advance, the ordnance dropped by the medium bombers was 260lb fragmentation bombs. 'Great fountains of earth were thrown up producing dust that mixed with smoke [that] gradually obscured the view, but rolling blasts continued and we could barely hear ourselves speak.'

One of the German soldiers on the receiving end of the bombing was *Oberleutnant* Bandomir, a company commander dug in near the farming hamlet of Le Mesnil Frémentel, which was to be attacked later in the morning by 8 RB:

Oberleutnant Bandomir, wearing a helmet, pointing out his defences to *Oberst* von Luck prior to GOODWOOD.

Shortly after dawn on 18 July, I returned to my CP shelter, then prepared my breakfast. Around 6.00am, we heard planes flying over the sea. We clearly saw squadrons of bombers flying high in the sky. They are innumerable and are flying straight towards us. At first we thought that they are going to bomb German cities and that makes us very unhappy. But, suddenly, the first wave of planes dropped their bombs, then the second, they fell in the Emieville sector, further east. Our Flak, installed around Caen, opens fire and shoots down a bomber. That's when the gates of hell open.

The … bombardment from the aviation and then the artillery will last almost two hours. We are then unresponsive, we sit and wait for death, unable to do anything. It seems that the attack was aimed at destroying our heavy weapons. I then see a well-buried 10.5 cm anti-tank piece [*sic* self-propelled gun of 4/200 *Sturmgeschütz*] which is destroyed by a direct hit. Thanks to our solid and deep trenches, well covered, human losses are relatively low. However, this demonstration of force of opposing air superiority is very strong because it shows our weakness. The psychological effect of the bombing is clear. The aerial bombardment was followed by several hours of barrage from naval artillery and land-based units. Shells plough every square metre of land in our area. My CP is hit two or three times but it holds up. Even animals are stressed by the bombing. A wild rabbit then jumps inside our bunker and climbs into my arms. Almost petrified, he drinks coffee from my cup (!) and makes a hole with his teeth in my jacket sleeve …

Sturman Jospueit recalled the damage done by the bombing:

> The shrapnel had destroyed all the equipment which lack of space in the trench had forced us to leave outside. Our losses included our machine gun and barrel case, our personal equipment, tarpaulins, gas masks, and pistols.[2]

The bombing destroyed and damaged much of the German armour deployed forward to support the 16th Luftwaffe *Felddivision* and neutralised most of the remainder for the crucial initial period of the British advance. They included the Tigers of *Leutnant* von Rosen's No. 3 Company 503 *Schwere* Panzer Battalion, located in Manderville just 3 miles behind the front line.

The medium and heavy guns of the AGRAs joined the cacophony with counter-battery fire but it was the field guns that all ranks of the armoured battlegroups were waiting for:

> The roar of artillery was terrific on either flank, and a small per centage of misdirected friendly shells fell in amongst us, as well as a few which were definitely hostile. We eagerly awaited the sound of the 25-pdrs starting up, which would be the signal to go forward. At last the creeping barrage began, and the mighty armoured steam-roller started on its task to flatten out all before it.[3]

Thanks to the shells dropping short, 3 RTR regimental group set off shortly after 0700 hours 'in some disorder' to the start line via the minefield gaps No. 17

One of von Rosen's 56-ton Tigers overturned by the bombing.

8 RB's M5 or M9 half-tracks photographed going through one of the minefield lanes. The 6ft pickets mark the edge of the lane and note the smoke and dust.

and 18, with dust turning the daylight into an orange gloom. Major Close recalled that A Squadron's column 'moved slowly down to the minefield nose to tail, closely followed by the carrier platoon of G Company … As expected, we had no difficulties in negotiating the taped lanes through the minefield although one of the carriers did stray out of its lane and lost a track on a mine.' He continued:

> Sure enough, as we formed up on the start line as the light came, my fears were well founded; a complete regimental group concentrated in such a small area must have presented a considerable target to the enemy. Strangely enough, no enemy fire was directed at us at that time and we busied ourselves with last-minute battle preparations.

Other than the carrier platoons, the rest of G Company was at the rear of the RTR's box formation:

> We had orders to form up in three waves; first wave – A and B Squadrons, second wave – RHQ, Recce Troop, Carrier Platoons, 1 Troop Flails, half Troop AVREs; third wave – Reserve Squadron, Motor Company, [H Battery] 13 Royal Horse Artillery.

Behind the RTR the F&F Yeo were on the move. Their war diary observed:

> At 0720 hrs the regiment moved off behind 3 R. Tks. through the start line [*sic* minefield] just west of Escoville 1171. The form-up at the start line was considerably hampered by a minefield laid by the Highland Div. which they had been unable to pick up except for four lanes [*sic*]. These lanes provided a further bottleneck [after the bridges] and had a serious effect on the outcome of the whole operations.[4]

This is an early indication of the problem with bottlenecks that was to blight the timely deployment of the armour on which the whole operation depended.

The Advance Begins

At 0745 hours the barrage started to creep forward at a rate of 5mph, with the leading squadrons, A right and B left, attempting to keep 100 yards behind it. The scene that greeted the armoured column is described by the divisional historian, who was impressed by the effect of the opening firing on the ground, if not the enemy:

> After the barrage had ceased and the last of the bombers had departed there lay before us a waste land, 5 miles of utter devastation; the trees blasted, the buildings shattered, the air foul with the death-stench of cattle and horses. There is a certain intangible nightmarish quality about such scenes, for the acrid pungency of high explosive tends to dissipate the sweeter emanations of the morning air, and the smoke and dust of battle to belie the serenity of a summer sky. Yet were the evidences of reality strong, for beyond and beside these fated acres the enemy waited behind his guns, and even within the dead land itself he began to emerge blinkingly from his foxholes ready to sell his life dearly to our advancing troops.[5]

The frontage on which 3 RTR was to advance through was christened the 'Tank Alley' by the riflemen. Initially just over 1,500 yards wide, as it was constrained by the village of Cuverville, the first objective of 159 Brigade, on the right and the villages that lay at the foot of the Bois de Bavent on the left, which were to be taken by the 3rd Division. However, 2 miles into the advance it narrowed to a choke point of just 1,200 yards between Lirose and Démouville, where the Caen–Troan railway line lay across the line of advance. Major Close described the advance down Tank Alley:

> We roared on through boiling clouds of dirt and fumes, thirty-eight Shermans doing their best to keep up with the rolling curtain of fire. I could vaguely see tanks on either side of me slowly picking their way through the ever-increasing number of bigger and bigger bomb craters. I only hoped that B Squadron on my left was also keeping up ...
>
> Dazed and shaken figures rose from the uncut corn and attempted to give themselves up to the leading tanks. When I waved them to the rear they stumbled off with their hands over their ears. Other Germans squatted in their foxholes staring stupidly, completely demoralised as we passed. Our infantry would collect them, or so we hoped. Sure enough, we could see [the G Company, 9 Platoon] carriers rounding them up and escorting them to the rear.
>
> As we moved slowly forwards, with engines whining as drivers changed down in order to bypass the still smoking bomb and shell craters, the barrage drew further away.

For the riflemen, particularly those in the carrier platoon advancing with the tanks, the experience of driving down Tank Alley early in the attack, threading

The 650 tons of bombs falling on Cagny had reduced the village to rubble.

their way through, was 'thrilling', even 'fun'. Fortunately, Major Becker's No. 1 Battery in Démouville had been badly hit by the bombing and or dispersed and did not have the impact on the RTR's advance it otherwise would have. Taking an hour to move down Tank Alley, the Shermans were making just 2mph and were, consequently, losing the barrage causing the CRA to order the two optional dwells short of the railway line. However, the Germans were still reeling from the bombing and the shelling, as recounted in 8 RB's battalion's report:

> During the advance of the first half of the barrage i.e. as far as the railway CAEN – TROAN, the Motor Companies took a more or less active part in collecting prisoners from 16 G.A.F. Fd Div who were doing their best to give themselves up after our bombardment. One carrier of H Coy which was left to guard a small batch of prisoners near DEMOUVILLE 1067 noticed considerable enemy activity in a wood nearby and fired a few shots to keep them quiet; 75 Germans thereupon came out of the wood, laid down their arms and surrendered.[6]

Also on the right, in cover south-east of Cuverville, the tanks noticed signs of enemy recovery. Over the radio 1 Troop received orders: 'Anti-tank guns firing from the orchard area; use your machine-guns to keep their heads down.' In one case a race to manhandle a Pak 40 through 90 degrees to bear on the Shermans and a similar turret traverse was won, with a HE round demolishing the anti-tank gun.[7] About a platoon's worth of infantry emerged to surrender.

One of 23 Hussars' Shermans photographed early in the battle amidst the smoke-induced gloom.

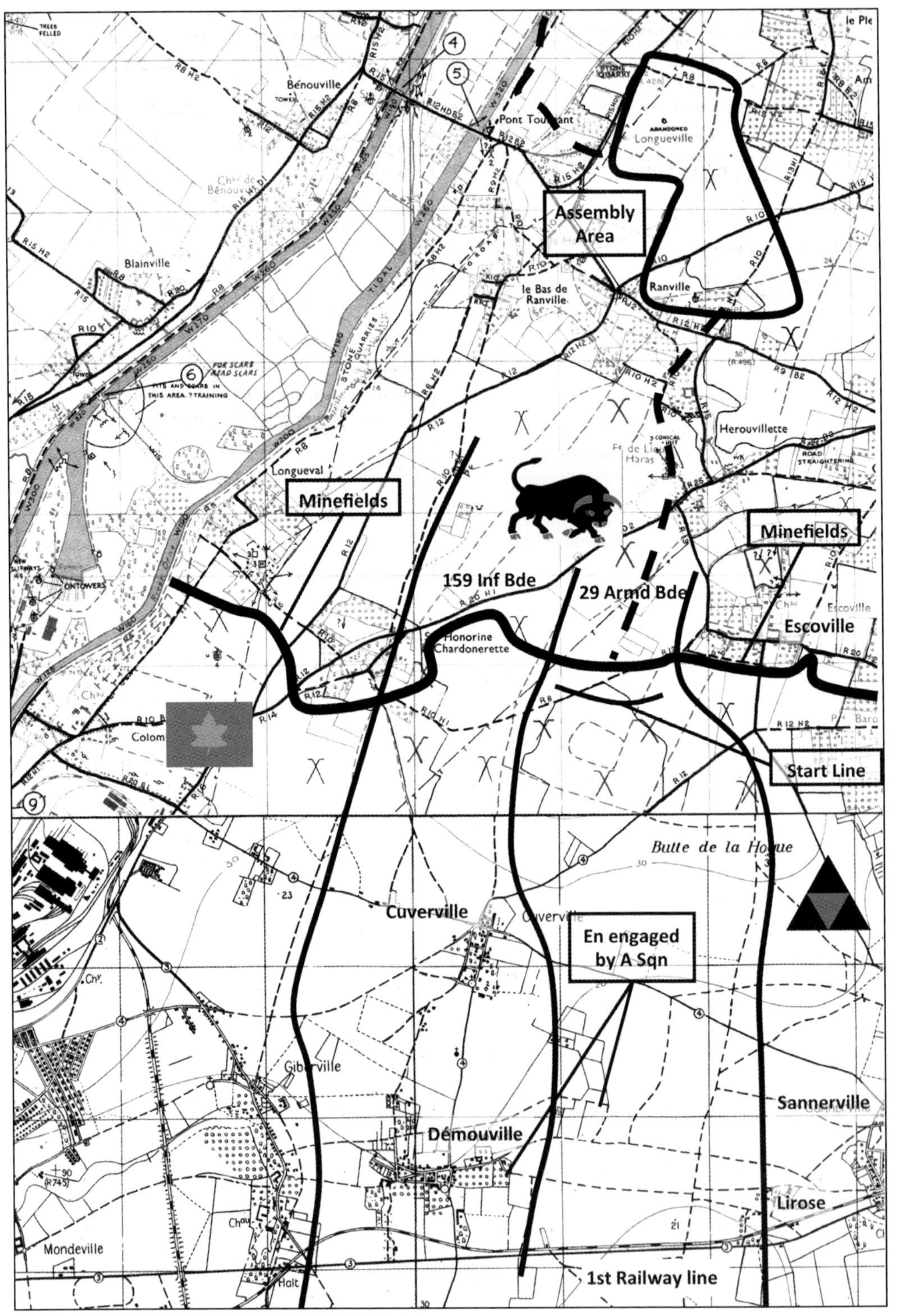

The advance to the first obstacle.

G Company had an unusual encounter at the rear of the RTR's column:

At first we saw little sign of the enemy, and the few Germans we did see appeared completely dazed as they came in and gave themselves up. A German armoured car appeared from some woods on our right and made a dash for it right through the middle of us, unfortunately getting away with it, as the only tanks which could fire with any degree of safety were those in front and on either flank.

The First Obstacle

The leading tanks reached the Caen–Troan railway at 0845 hours already half an hour behind schedule, forcing the division's CRA to order the first of two optional pauses beyond the railway line that had been built into the artillery fire plan. Consequently, the barrage again 'dwelled' beyond the Caen Troran railway, on report line MATILDA for fifteen minutes, while the armour closed up and crossed the First Obstacle.

From air photographs and maps, intelligence believed that the railway represented an insignificant obstacle. In this they could claim to be correct as a single tank or wheeled vehicle, the latter in places, could make its way carefully across the single set of tracks.[8] However, with a mass of armour having to cross, the rail lines made a significant obstacle at the mouth of the choke point. To produce a level track for the trains, the builders had created cuttings and embankments in the undulating ground that were between 2–6ft in depth/height. With hedges and ditches at some points, places where the tanks could cross were quickly found but it was altogether more difficult to find places for wheeled and half-track vehicles to cross. The services of the sapper demolition NCOs were called forward from the accompanying AVREs of 26 Assault Squadron to blow the rails

A Sherman negotiating the drop into a shallow cutting on the first railway obstacle. The sappers on the right have blown the railway lines and are preparing to widen the gap. The white building in the background is seen in the image overleaf.

The building by the railway crossing between Démouville and Baneville.

The three railways encountered by 8 RB during the first day of Operation GOODWOOD are a source of much confusion as different names are used in various accounts.

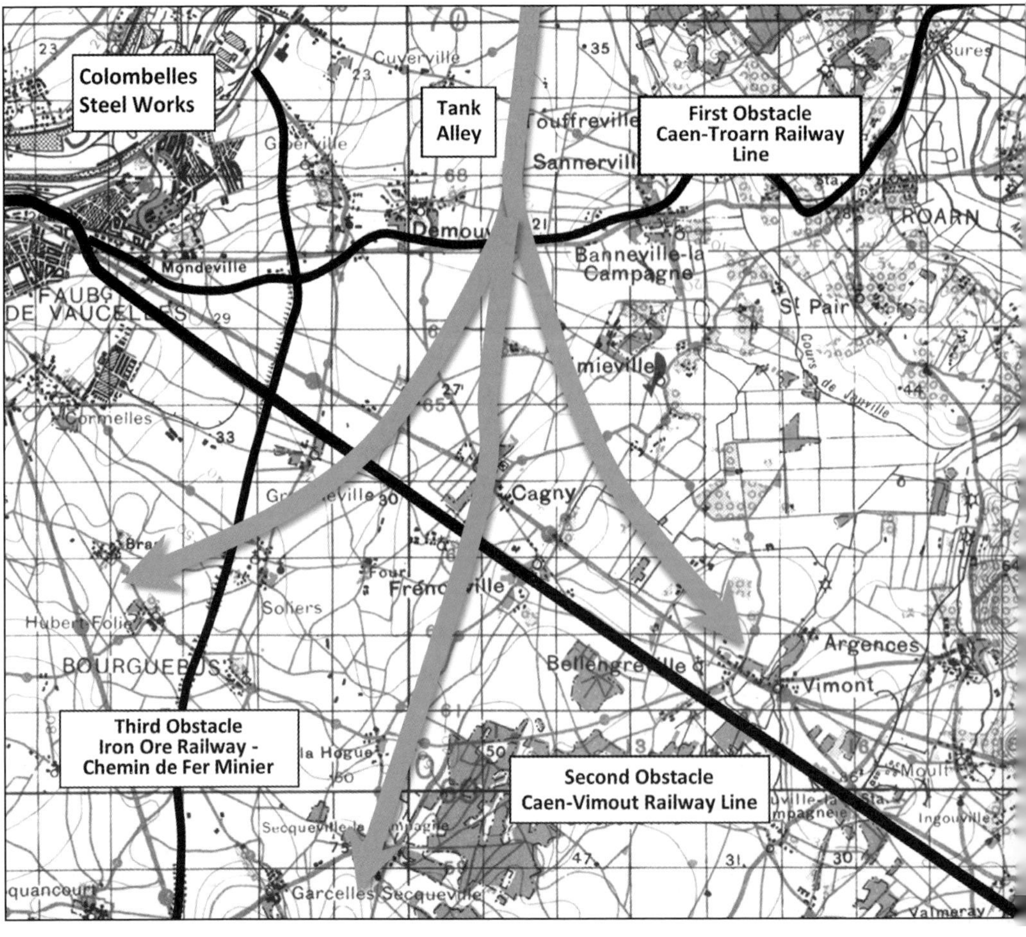

along with bulldozers to create ramps to speed up the crossing. By the time this had been achieved, significant traffic jams had built up and it was only at about 0900 hours that the leading squadrons set off south-west following the second phase of the bombardment.

The Advance Resumed

It was only several hundred yards before the tanks of 3 RTR encountered their next but less significant obstacle, the Sailes hedge, which again stretched across the line of advance. It was embanked and dense, with stout trees, through which 3 RTR struggled to find crossing points to the west of the Démouville–Cagny road. The tanks now came under artillery fire from the self-propelled 105mm guns of Major Becker's No. 4 Battery positioned around the hamlet of Le Mesnil Frémentel, but once through the hedge they were also engaged by SP anti-tank guns and some towed Pak 40s:

> As we moved up a slight slope towards the village of Le Mesnil Frémentel odd rounds of solid shot started whistling by and it was obviously the first signs of organised resistance. Suddenly a Sherman on my left rolled to a halt belching smoke. Immediately, every tank turned its guns on the houses in the village from where the shot had come.
>
> I could see considerable activity among the trees in the orchard and flashes from anti-tank guns in the line of hedgerows around the orchard. Two more

Major Becker, commander of *Sturmgeschütz* Battalion 200, was mobile on the battlefield controlling the withdrawal of his batteries.

tanks from the Recce troop were hit before the concentrated fire from the Shermans quickly dealt with our attackers, knocking out three or four guns and their crews.

The fire on 3 RTR was from I/125 *Panzergrenadier* Regiment, who had deployed out in the fields beyond Le Mesnil Frémentel and had recovered from the neutralisation by the bombers and artillery. Leutnant Bandomir's No. 3 Company headquarters was dug in in the hedgerows that surrounded the hamlet, while the battalion headquarters was sited in a bunker dug in the centre of the farm complex's courtyard. It was, however, not the job of 3 RTR to secure the hamlet. Instead, the tanks raced on past it to the west, taking several more tank casualties in the process. With the danger of being enveloped, No. 4 Battery disengaged and withdrew south almost certainly to an intermediate position in the cover of the bombed ruins of Cagny, while 3 RTR continued its south-westerly trend towards Grentheville.

While the RTR and G Company pressed on, the F&F Yeo along with F Company had made it through Tank Alley and across the first railway line and, with the ground opening out, were coming up to the left of 3 RTR. Both tanks and infantry were to the east of Le Mesnil Frémentel and had closed up to the field gun barrage that was almost at its maximum range. The yeomanry's war diary noted that so far all was going to plan:

> The regiment was covering a frontage of about 700 yds. The two leading waves encountered little opposition but much enemy equipment was seen to be abandoned and many prisoners attempted to give themselves up; as there was no time to collect these they were left to be mopped up by subsequent infantry.
>
> The two leading waves reached the line of the railway Caen-Vimont 1461 without suffering any serious casualties.

The reserve, C Squadron, was, however, still to the rear passing Le Mesnil Frémentel when they were 'strongly engaged by anti-tank guns from the area of Cagny 1164 and the woods to the northeast of it':

> The first tank to be hit was that of Major C. Nicholls, and almost immediately Capt. J.E.F. Miller's tank was destroyed. The rear troop of 'B' sqn., which was doing flank protection, was also involved and in all some 12 tanks were destroyed in this area before the situation was in hand.

Lieutenant Workman of 1 Troop C Squadron witnessed the carnage:

> It was a summer's day: there was blazing tanks, wounded men running around – people who had got out of a tank that had been brewed upcoming staggering back towards us. It was an awful scene, people staggering past covered in blood, being burned. We were pretty sure it was 88mm guns, it wasn't dug-in tanks. It was the most feared thing – and the first thing ever reported on the wireless and they'd give a map reference – that went straight

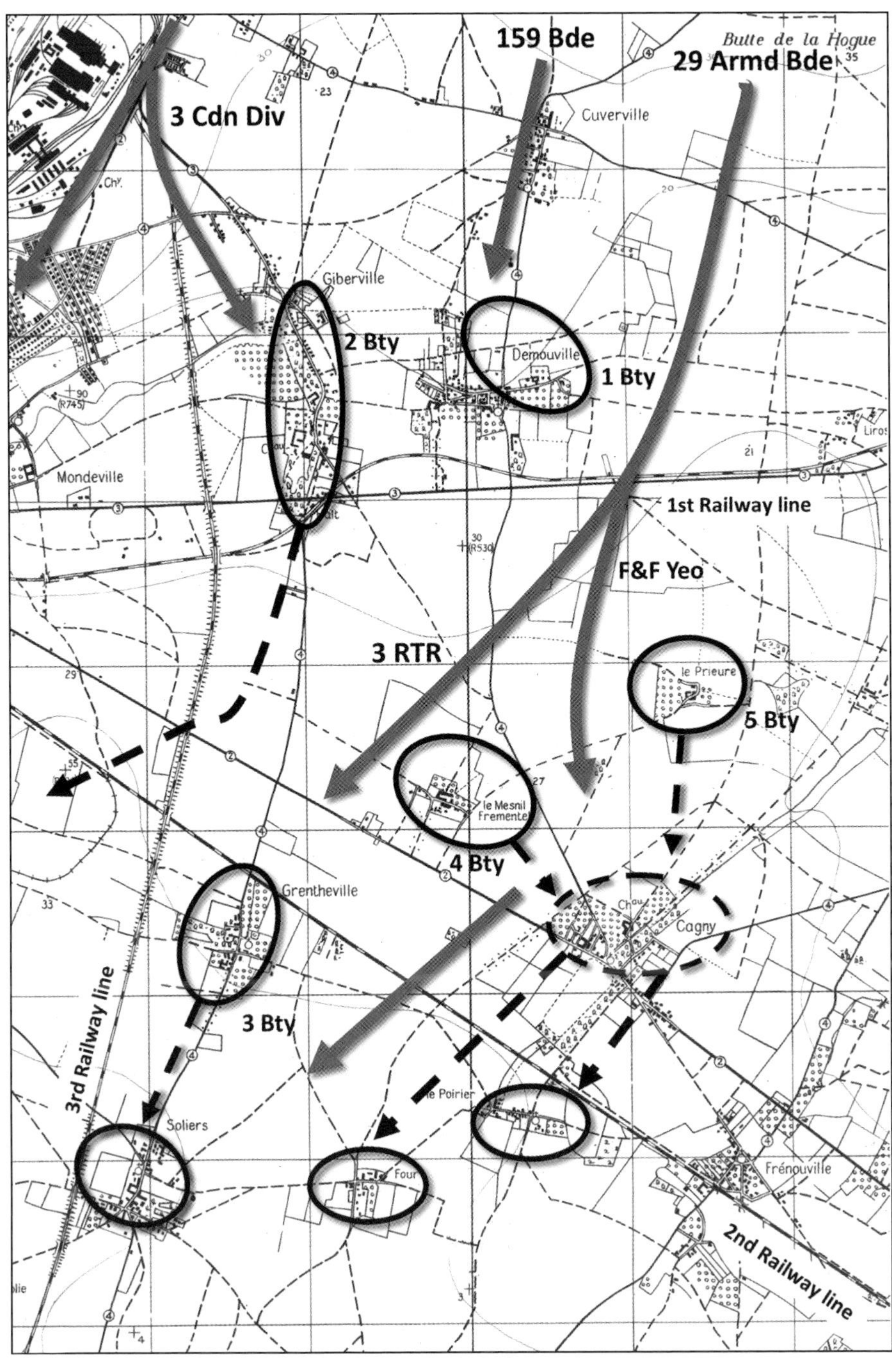

The withdrawal of *Sturmgeschütz* Battalion 200's batteries.

down on your map. Tank after tank after tank just going up! … We weren't
hit, but the people to my right were all being hit – several were killed.[9]

The yeomanry's CO believed that the fire was coming from Cagny and was in the
process of calling up F Company to clear the ruins of the village, when General
Roberts arrived and climbed aboard Colonel Scott's, tank where he was already in
conversation with Brigadier Harvey. The general late wrote that:

> The Fife and Forfar squadron was burning in the field ahead and there was
> quite a lot of mortaring coming down in the area of the village in front. As I
> arrived, I heard Alec Scott of the Fife and Forfar tell the brigadier that he
> was ordering his motor company of 8th Rifle Brigade to attack Cagny. He
> was pretty certain that the fire had come from there. But you will remember
> that I had specifically asked to be relieved from having to take Cagny. We
> had 23rd Hussars in reserve looking after the left flank until the Guards
> Armoured arrived very shortly. So I told Harvey to cancel any attack by the
> motor company and to concentrate on getting forward onto the Bourguébus
> ridge. Had I not told Harvey to cancel the attack, the company would have
> virtually walked straight into Cagny. In the event it was to take the Guards
> the rest of the day to capture the village after a considerable battle. How
> unfortunate it was that I arrived just in time to cancel Scott's attack.

The company's historian noted that: 'There was not much for F Company to do
during the day, because this was largely a tank v. tank battle, but what we *did* have
to do was to dig and lie low from time to time, as Black [armour-piercing rounds]
from both sides came bounding into our midst.'

The situation could have been worse as following the bombing *Leutnant* von
Rosen had managed to get six of his Tigers running and led them to attack
29 Armoured Brigade's left flank. However, as they approached 1,000 yards from
Cagny, two Tigers were knocked out in quick succession by shots through their
frontal armour. Believing that the British had a new weapon, the Tigers promptly
withdrew. The source of these shots has long been debated. One candidate is an
88mm anti-aircraft battery that had survived the bombing of Cagny on its
northern edge, another is Becker's Pax 40s that had withdrawn to the village and,
of course, a Firefly's 17-pounder discarding sabot round. To support this latter
possibility, the 23 Hussars claim to have knocked out a Tiger during their
advance south of line MATILDA. If these powerful German tanks had been able
to press home their attack into the flank of the brigade the leading tanks could
have been cut off and destroyed at leisure by the Germans.

Rifleman Gillate of G Company's 9 (Carrier) Platoon, who was with the RTR
further to the right, recalled that matters were scarcely better on the other flank:

> Things became very hairy indeed. They picked off our tanks like boxes of
> matches. They were brewing up all round us, it was absolutely terrible to see
> and it was equally terrifying to hear because they were using solid shot,
> armour-piercing shot. This made a sort of noise that high-explosive shells

Leutnant von Rossen commander of 3/503 *Schwere* Panzer Battalion.

did not make, it was like a clapping sound, perhaps why they call it a clap of thunder. It was not a bang, an explosive bang, it was a twanging sort of bang. The only way in which I could really describe it is that if you were to put your ear fast against a solid metal bath and somebody took a sledgehammer on the other side of the bath and hit the bath very hard, your ear would practically drop off. That is the kind of noise that this made, it was absolutely frightening. The most frightening aspect of course being: which is the one with your number on it?[10]

Casualties among the leading armoured regiments mounted inexorably.

Battalion Headquarters in Action

At the rear of 29 Armoured Brigade's column was Lieutenant Colonel Hunter's headquarters of 8 RB with its usual gaggle of attachments. The brigade's Op O No. 3 detailed the battalion's grouping and task:

8 RB with under command 119/75 Atk Bty, for mov only 1 troop 612 Fd Sqn and after completion of breaching one sqn 22 D [Flails] will seize and hold LE MESNIL FREMANTEL in order to secure LEFT FLANK of bde agst any attack from the EAST or SE, until relieved by another fmn.[11]

It was anticipated that not only would it be some time before the Guards Armoured Division came up on the left, where von Rosen's Tigers advanced from, but also as 159 Brigade was advancing and clearing Cuverville and Démouville on foot, the right flank would also be open for a while. Once Le Mesnil Frémentel was cleared, the plan was for the twelve M10 Tank Destroyers of

A command half-track, probably E Company's, in a column of vehicles negotiating a minefield gap.

119 Battery to be positioned around the hamlet and paddocks to the north forming a 'firm base'.

To secure even a small farming hamlet required riflemen's boots on the ground but with only E Company's headquarters and two machine gun platoons, and with drivers being required for numerous vehicles, infantrymen were in short supply. Consequently, Colonel Hunter took all the riflemen who could be spared from across Headquarter Company and formed them into a platoon-sized force under E Company's Major Rowan. As suggested by air photography, they were mounted for mobility in a Dingo scout car and three half-tracks. Fire support was to be provided by the flail tanks of a squadron of 22 Dragoons but in the event, Shermans of 23 Hussars were moving south on the axis of the Démouville–Cagny road and were available to provide close support.

Having crossed the start line at 0915 hours and made their way over the railway line, Colonel Hunter's command was ordered to begin their part in the operation

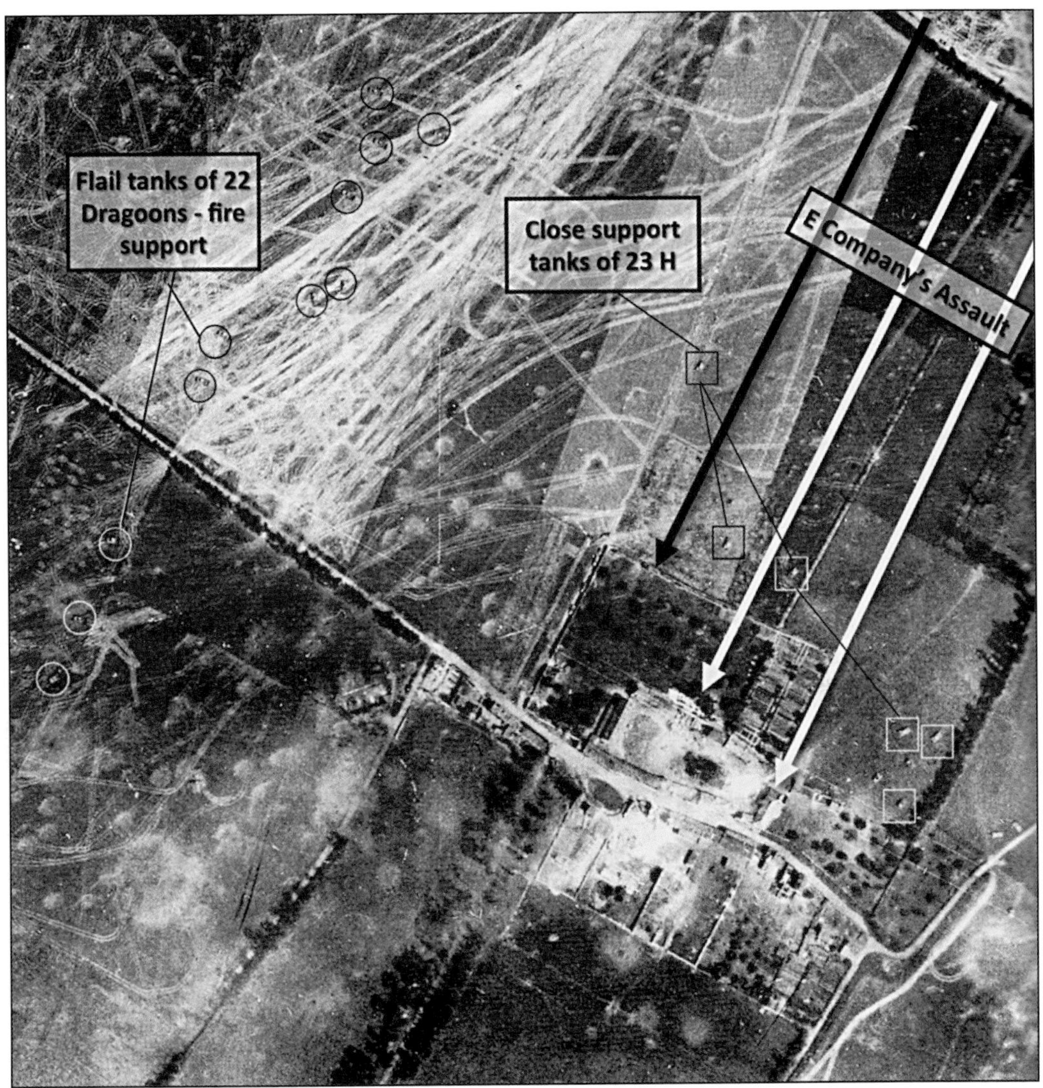

8 RB's attack on Le Mesnil Frémentel.

at 0955 hours. Five minutes later they reported crossing the railway line and little over thirty-five minutes later, at 1038 hours, they were recorded as being in action, in and around the hedgerows north of Le Mesnil Frémentel, which had been bypassed by the armoured regiments. Before tackling the hamlet itself, the riflemen's first task was to clear No. 3 Company I/125 *Panzergrenadier* Regiment from the hedges but with the fire and menacing assistance of the Shermans and flails, the demoralised *panzergrenadiers* were quickly winkled out of their bunkers and trenches with minimal resistance and sent off to the rear.

The assault on the hamlet itself was from the north, with flails providing fire support from the west. The British armour was largely unopposed as by this time Becker's No. 4 Company had withdrawn south, leaving one gun destroyed. While Colonel Hunter commanded the attack as a whole from the area of the Sailes

Hedge, the infantry assault was led by Major Rowan across open ground to the orchard to the north of the farm buildings. In this they had the close support of two troops of Shermans of 23 Hussars. Consequently, the riflemen's advance was quick and methodical, with any defenders showing signs of resistance being dealt with by the supporting tanks, which closed in on the buildings as E Company advanced. However, once inside the orchard and the hamlet it was down to the riflemen to clear the headquarters of I/125 *Panzergrenadier* Regiment with rifle, bayonet and grenade. Again, the enemy were largely dead, wounded or cowering in their trenches. *Oberleutnant* Bandomir, commanding No. 3 Company, had survived the bombardment in his deeply dug command post beyond the stone wall surrounding the orchard. With Becker's guns having withdrawn and no anti-tank defence of his own, Bandomir provides a German perspective of 8 RB's attack as he emerged to fight:

> Surprised to find ourselves still alive, we looked over the edge of our trench. I saw many tanks approaching. I scaled a wall, hoping to find some of my company ... So we decided to leave the village and try to head south to rejoin the regiment.
>
> The tanks fired machine guns on us and we walk along the wall surrounding the farm in the hope of finding a few men from my company. But no one survived in our forward positions. I then jump through a gap in the wall to reach our CP in the farmyard. I retreat immediately because a [Sherman] tank is already in position on the roof of the CP. We decide to leave the farm by running south to reach our regiment's headquarters in Frénouville. We reach a sunken path lined with trees, there we find numerous bodies of killed

The Sherman Crab or flail had a very effective 75mm HE round and two Browning machine guns that together provided significant fire support to 8 RB's assault on Le Mesnil Frémentel.

or wounded men. Those who are still able to fight joined us when we left the farm. We hide in a wheat field to wait out the night. But, still, the tank machine guns are mowing down more and more of our men with their fire. I then give the order of surrender to those who survived. It was around noon that the war ended for me and the survivors of the 3rd Company. The situation leaves me no possibility of escape.

Without anti-tank weapons and with numbers that had become too thin, it was impossible for me to command my company, especially since my lines of communications were destroyed. I am then a prisoner.

Oberleutnant Bandomir concluded: '18 July 1944 was a hot and sunny day in Normandy. But for me and my company it was the most dismal and depressing day in all our operations, because we were powerless to do anything.' Captain Finnes, the battalion's adjutant, joined in rounding up prisoners:

Armed only with a pistol I decided to investigate a number of trenches near battalion HQ. My German was very limited, but waving my pistol in what I hoped was a menacing manner and shouting '*Aus, aus*' seemed to have the desired effect and a number of bedraggled Germans emerged. They did not seem keen to fight. I suppose that the bombardment and the speed of our advance had undermined their morale. But I remember thinking how odd it was that on a fine, hot July morning almost all of them seemed to be wearing long trench coats. By 11.30 am we had cleared the whole area and taken 134 prisoners from 1st Battalion 125th Panzer Grenadier Regiment. We sent them off with a small escort heading north.[12]

At 1145 hours, Colonel Hunter reported that he had positioned his slim infantry resources and the guns of 119 Anti-Tank battery in defence around Le Mesnil Frémentel. With 8 RB having spent most of the rest of the day 'continuing to round up Germans from their trenches, bunkers and other hiding places', I/125 *Panzergrenadier* Regiment had virtually ceased to exist.

Opposition Mounts

By 1100 hours, the leading squadrons of the F&F Yeo and 3 RTR had reached the second railway line, which being dual track was altogether more difficult to cross. At the same time, while Becker's companies were falling back to the villages to the south, reinforcements were arriving on the Bourguébus Ridge to block the British advance. On hand were the corps' 88mm Pak 43 guns from 1039 *Panzer-jäger Abteilung*, which were already deploying, to be shortly joined by elements of the 1st *Leibstandarte* SS Panzer Division that had been released from the Orne further west. First to arrive was the recce battalion, which unlike British recce regiments was a potent fighting force that bolstered the flagging resistance of the 21st Panzer Division. Panzers were also on their way, as were *panzer-grenadiers*, but it would be some time before most of them arrived on the Bourguébus Ridge. Meanwhile, a squadron of 23 Hussars was engaging panzers approaching from the area of Frénouville to the east. The situation that had hitherto been broadly favourable had taken a definite turn for the worst. The distinctive sound of armour-piercing rounds passing overhead now greeted the riflemen in their carriers. Fortunately, low in the standing crops that had not been beaten down by bombing and shelling, they were not the prime target; with the slap of steel on armour, the Shermans started to be hit and burned with oily black smoke.

As the 3 RTR battlegroup crossed the Caen–Vimont line and approached Grentheville, the hitherto inactive crews of 14 *Nebelwerfer* Regiment's launchers recovered from the barrage and creeping bombardment. Major Bell recalled:

> Here a 'Moaning Minnie' opened up just in front of us, but before the last of its six barrels had been emptied the turrets of a dozen Shermans swung round and blew it and the crew to pieces – the best thing we had ever seen happen to this diabolical weapon.

The next challenge for 3 RTR, less A Squadron, was to advance through the narrow gap of just 200 yards between the iron ore railway and the wood and orchard west of the Grentheville Monastery. Colonel Silvertop ordered G Company to secure their passage by clearing the edge of the woods. This task was given to a platoon that was warned not to get drawn into clearing the whole area, as the company would be needed for the next phase of the advance towards the villages atop the Bourguébus Ridge. However, when the tanks moved on to the west of the iron ore railway, the half-tracks came under fire, leaving the re-embussed G Company

> stuck out in front with no one upon either flank, and we were ordered to wait here until the opposition encountered on our left had been cleared up.

Major Bill Close MC, commander of A Squadron, 3 RTR.

Shelling became rather unpleasant owing to our being in full observation of the enemy, and there was very little we could do about it. It was no use moving because the shells just kept on following us around.

Sergeant Hicks with H Company was equally exposed:

Heavy mortaring from their six-barrelled rocket mortars and artillery fire descended upon us. Twice shells exploded immediately in front of my own Bren gun carrier which lifted it up as we ran into the explosions. Another yard and we would all have been killed but we were lucky and on we went, a little deafened and even dazed for a moment, but unscathed and mechanically undamaged but with a few dents.

German anti-tank guns along the Caen–Argentan railway embankment and in other well-prepared positions began to take their toll of our Sherman tanks and the advance ground to a halt. There followed a long and bitter tank battle with the German Panther tanks and there was little we infantry could do about it stuck out in the open country soaking up all that the Germans could throw at us.

We dug deep slit trenches and watched this awesome panorama where tank after tank was hit and exploded into flames, sending black smoke belching up into the hot clear blue sky of the summer's day ...[13]

Despite repeated and increasingly unrealistic orders for the armoured regiments to advance, there was little scope for 8 RB's motor companies to play a part in a major tank *vs* tank battle. By 1230 hours the advance of 29 Armoured Brigade had stalled with tanks burning. As the de-horsed tank crews made their way back, G Company's carrier platoon, with the other motor company's carriers presumably similarly employed, collected wounded and burned men and took them back to an ambulance exchange point or an RAP near the minefield.

With the armoured battle raging, G Company left its scout platoon in the western end of Grenthenville and tucked themselves away in the cover below the iron ore railway embankment near Colonel Silvertop's headquarters. Major Close recalled that:

RHQ down by the embankment was being rather heavily bombarded by artillery fire and it was not until one of the Rifle Brigade sergeants wandering along the embankment found a German observer well dug into the side of the railway – he summarily dealt with him – that things became a little quieter.

During the course of the day other enemy observers that the advance had rolled over were found and killed or taken prisoner, particularly in the areas that had not been bombed.

While G Company was in the dubious cover under the railway embankment alongside RHQ, Colonel Silvertop considered his next move onto the ridge, which was dominated by the villages of Hubert-Folie and Bras. Lieutenant

Stileman, the commander of 10 Platoon, was dismounted and probably hovering around RHQ trying to find out what was happening:

It so happened that my platoon command half-track was almost next to the gallant colonel's tank. I saw this finger beckoning me over. A pair of steely eyes studied me. 'I've got to find out whether Hubert-Folie is occupied or not', he said. 'Jolly good idea, sir', I said, or something equally fatuous, 'How are you going to do it?' The eyes still studied me, 'You're going to', he said. Like all good military plans this one was delightfully simple. I was to leave my half-track, take two carriers from the carrier platoon, work my way up through the corn alongside the railway embankment, then head for Hubert-Folie church. On getting there I was to drive from left to right straight through the middle of the village. If we emerged at the far end the chances were that the village was not occupied. If we didn't, it probably was. It was

A carrier leads a column of half-tracks past a knocked out Sd.kfz.251.

not one of those orders groups at which 'Any questions?' are called for! Bill Close has been kind enough to describe me earlier as 'a florid cockerel'. At the prospect of this charade I must confess that I felt more like an anaemic broiler![14]

This, however, was not going to be an ill-considered dash:

Now, it so happened that Noel Bell produced a marvellous air photograph of the village of Hubert-Folie and the surrounding area. Well of course this meant that I didn't have to fuss around with maps. And also at this moment [Major] Bill Smyth-Osborne, who commanded H Battery of 13th RHA, offered his services. And he decided to put a heavy artillery concentration down on Hubert-Folie as we approached to the village. And the last shell was to be a [white] phosphorous one, and this was the signal for us to start on our journey down the village.

The carriers duly approached to within 100 yards of Hubert-Folie under cover of a shoot by the Sextons of 13 RHA:

We made our way up towards the east side of the village, and sure enough, the smoke round landed smack on the church tower. That was the signal to go. I gave the order to my driver, Rifleman Butler – and we didn't loiter! We went flat out straight down the road through the middle of the village, between the orchards and the houses, which are now much as they were, like a rat up a drainpipe. Then out the far end, we swung right and came straight back down the slope. I then reported that we hadn't seen any enemy and that since I had returned I didn't think it could be occupied. But how wrong can one be, because we discovered shortly afterwards that the place was simply groaning with Krauts.

At the time this recce took place it is almost certain that the *Leibstandarte*'s recce battalion was still arriving on the ridge and villages, and that the SS infantry that eventually occupied the village would mostly only do so as darkness fell later that day.

Both armoured regiments having been halted with only the odd troop slipping through on to the ridge, 23 Hussars came forward and resumed the attack but to no avail, as by now the *Leibstandarte*'s *Sturmgeschütz* battalion had arrived on the ridge and added to the volume of anti-tank fire being directed at the British armour. The brigade fell back behind Soliers and Four, both of which remained in enemy hands. The tanks of 29 Armoured Brigade had advanced some 12,000 yards but the German defence in depth, which was much greater than intelligence estimated, had absorbed the brigade's thrust. Meanwhile, after much delay the Guards Armoured Division were in action to the left at Cagny, in what was their first battle.

As the *Leibstandarte*'s panzers arrived during the afternoon, the Germans mounted the counter-attacks that their doctrine required throughout the

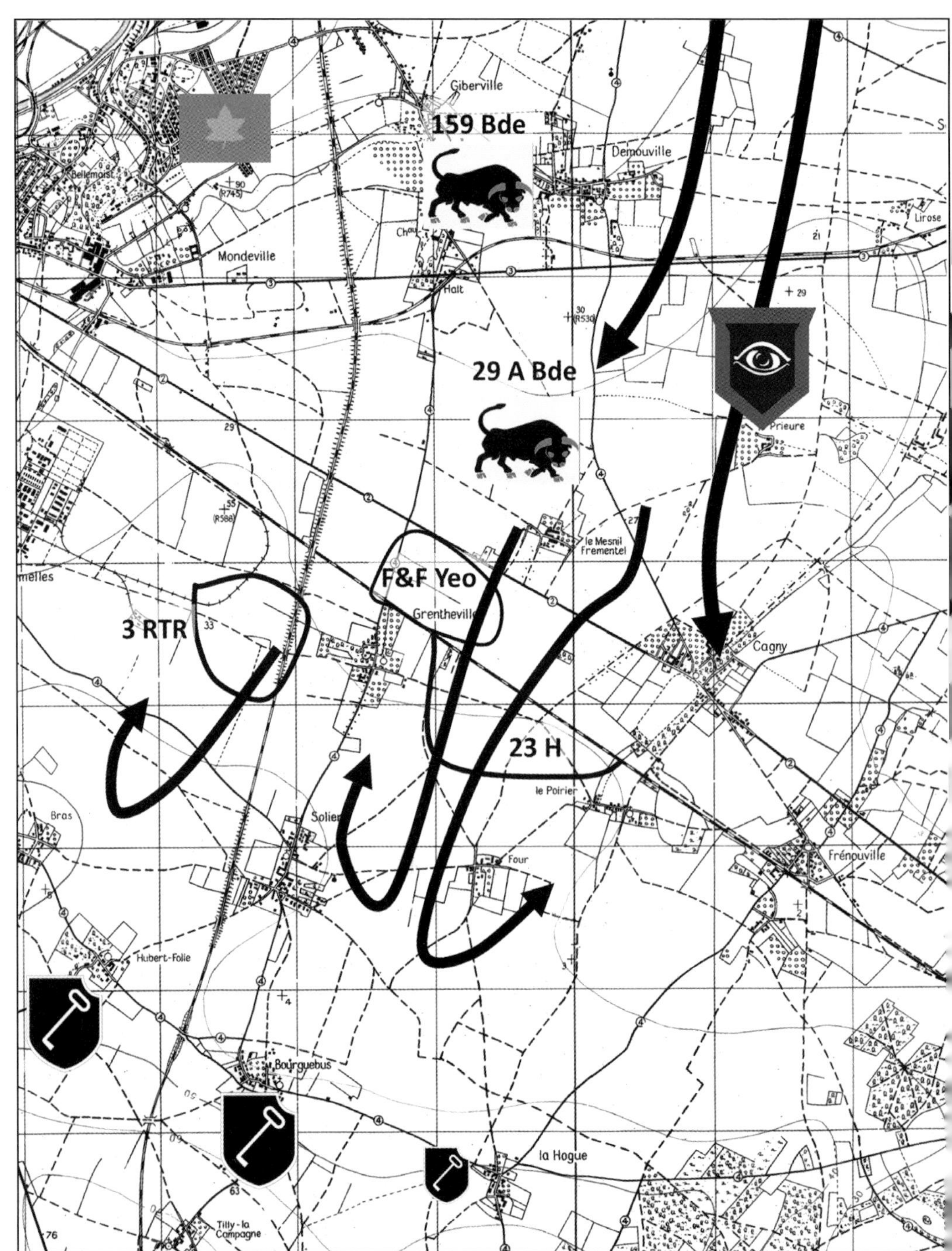

The high-water mark of 29 Armoured Brigade's advance on 18 July and overnight positions.

afternoon, no fewer than five of them. They were, however, largely broken up by the fighter-bombers of 83 Group's twenty-five squadrons, as the artillery, less 13 RHA, was now either out of range, at their extreme range, or moving forward.

Major Bell provides a glimpse of what it was like under increasing fire from the ridge:

> ... things rapidly became very unpleasant for us. Armour-piercing shells began coming in from all directions and we were unable for a time to pinpoint where any of them were coming from and tanks of the 3rd R.T.R. began 'brewing up'. Then on our left, Panthers appeared and the fun really began. The tanks had pulled back to positions from where they could engage the enemy a little more safely, and I found that my half-track and a section of carriers were stuck out in front of them all. The best we could do was to move back to a hedge behind which we could at least get a little cover from view, although none from fire. Unfortunately, the carrier section suffered heavy casualties, and it was only a very small percentage of their strength which managed to join us at the hedge ...
>
> We seemed to lie behind that hedge for hours, imagining every moment to be our last ... During this period the Northants Yeomanry, whom we had been awaiting, arrived, but they did not appear to be quite in the picture, and it was not long before they too were suffering disastrous losses.
>
> Eventually we were able to pull back and rejoin the rest of the Company, who by this time were hard at work with their digging materials ...

Carriers of F Company 8 RB photographed during 18 July 1944.

Grentheville

Earlier G Company had deployed a platoon to secure passage of 3 RTR past the woods and orchards to the west of Grentheville and F Company had deployed 16 Platoon on a similar task to cover 23 Hussars past the east of the village. Fortunately, the guns of Major Becker's No. 3 Company that had been in and around Grentheville had finally withdrawn to the next village, Soliers. This left the village in the hands of the infantry of I/125 *Panzergrenadier* Regiment, or parts of it that had not been pushed south of the Vimont railway. Major Cunliffe's F Company, which had been following in the wake of F&F Yeo, was ordered to clear Grentheville and its monastery. Lieutenant Sedgwick's citation for the Military Cross provides an insight into the operation:

> On the evening of 18th July, 1944, F Company of a Territorial Battalion of the Regiment was ordered to clear the village of Grentheville. Lieut. Sedgwick commanded No. 6 Platoon, which was the leading platoon into the village. He showed great dash and determined leadership in overrunning two orchards full of the enemy and capturing twenty prisoners and twelve nebelwerfers.[15]

Lieutenant Sedgewick was supported by Lance Sergeant Triggs, who commanded a section of 5 (Scout) Platoon's carriers and earned the Military Medal:

> His section was attached to a platoon which had been ordered to carry out an attack on and clear the village of Grentheville. While taking up position covering the village, L./Sergt. Triggs had his carrier hit and destroyed. He immediately evacuated his wounded gunner, jumped into another carrier and continued with the action.
>
> During the clearing of the village L./Sergt. Triggs captured a 75-mm. gun and towed it behind his carrier.

Not only were *nebelwerfers* from 7 *Werferbrigade* captured but many of their crews and infantry from the 21st Panzer Division were as well.

The brigade war diary records: '8 RB found 24 *nebelwerfers* in GRENTHE-VILLE. They destroyed 23 and sent the 24th back to Int'.

Lieutenant Sedgewick's MC citation continued: 'Throughout the evening and the ensuing night Lieut. Sedgwick's platoon were under heavy artillery and mortar fire and he showed great coolness and courage, being a really good example to his men.' F Company remained in Grentheville '... where we spent what was probably the most unpleasant night of the entire campaign'.

As for the rest of 8 RB, the headquarters joined Brigadier Harvey's Tac HQ, with H Company dug in nearby as a blocking position with the M10 tank destroyers, while G Company covered the gap between the iron ore railway and Grentheville. Of this Major Bell wrote:

> Plans were then made for our night dispositions, and as a result we moved back behind a railway embankment, which must have saved many of our

Lieutenant Sedgwick receiving his Military Cross.

lives, as shortly after we had got into our new positions 88s opened up from the village of Bras and tanks began brewing up one after another. A very persistent 'Moaning Minnie' was also sending over its missiles, which were dropping too close to be pleasant. So after our great start, we had ended up the day a very depleted force, and it was clear that we would be unable to carry out our original intention.

To make matters worse, at 2330 hours the Luftwaffe launched one of their more significant bomber raids, scattering anti-personnel bombs across the GOOD-WOOD area. Fortunately, the riflemen were dug in and suffered less from bomb shrapnel than those sheltering under their tanks, although a number of the half-tracks were damaged.

The Vickers Machine Gun

The Vickers machine gun was adopted by the British Army in 1912, with two guns being issued to each infantry battalion. However, during the First World War its numbers increased dramatically, and specialist machine gun companies were formed as a part of the Machine Gun Corps, which was disbanded in 1922, and the guns returned to the infantry battalions.

In the early 1930s machine gun companies were reformed before, in 1936, four-line infantry regiments became specialised Infantry (Machine Gun) Battalions. They provided thirty-six Vickers guns at divisional level in infantry formations and provided companies of twelve guns for independent brigades and those infantry brigades in armoured formations. The three specialist regiments were the Middlesex, Cheshire and Northumberland Fusiliers. Amongst the numerous exceptions were airborne and motor battalions, who had their own MMG platoons. 8 RB held four Vickers in each of its two platoons (20 and 21).

The Vickers medium machine gun had undergone numerous changes to gun, tripod, ammunition and ancillaries in its long service prior to 1944, but its essence as a tripod-mounted, water-cooled and belt-fed weapon remained. It fired .303 ammunition from 250-round canvas belts.

The effective range was the same as the Bren gun, around 2,000 yards, but with the Mk VIIIz bombardment round indirect fire was possible out to 4,500 yards. It is unlikely that machine gunners in motor battalion's MMG platoons were fully competent in indirect fire.

Even though the Vickers gun could be kept in action by a crew of three, it needed a total of seven or eight men to keep it supplied with ammunition and to move it and its ancillaries away from its vehicle. This number included the driver of a Universal carrier, which each gun needed to transport it and the crew.

Chapter Seven

Operation GOODWOOD – Day Two

As far as 8 RB was concerned: 'When light came next morning, we did not know what the day would have in store for us. A small comfort was that it surely could not be worse than the previous one.' As recorded in the battalion's war diary, the battle was certainly not renewed with the coming of daylight on 19 July, as General Roberts reported to Corps Headquarters that he would need some hours to reorganise his division following the previous days' casualties:

> The next morning 23 H went forward to relieve 3 RTR who were keeping BRAS and HUBERT'S FOLIE [*sic*] under observation, but motor coys remained in posn and in fact the Bde virtually sat tight while 22 Armd Bde [7th Armoured Division] came up on our left with the objective of BOUGEBOUS and II Canadian Corps cleared the suburb of VACLELLES and CORMELLES 0565, thus securing our right flank.[1]

Throughout the morning F Company waited in Grentheville for relief in place and a move to join the rest of the brigade west of the iron ore railway. The SS infantry were active, as indicated by Corporal Fitzgerald's citation for the Military Medal:

> On the morning of 19th July, 1944, No. 6 Platoon, of F Company of the 8th Battalion of the Regiment, were holding the forward edge of the village of Grentheville, which had been captured the evening before.
>
> Cpl. Fitzgerald was in command of a leading section. A machine-gun 42 suddenly opened fire on the section and wounded one of Cpl. Fitzgerald's men who was forward of the section position.
>
> In full view of the enemy and showing complete disregard of his own safety, Cpl. Fitzgerald went forward and brought in the wounded man on his shoulder.

F Company, holding Grentheville overnight and into the morning, wrote in their account that: 'The enemy shelled and mortared us quite considerably, and by the time we had been relieved by some of 1st Battalion, The Rifle Brigade [7th Armoured Division], at midday on the 19th, we had suffered more casualties than we could afford.' Further back, Sergeant Hicks and H Company were under fire:

> It was a very frustrating morning as we were under direct enemy observation and, without orders to do anything positive, we just had to sit in the open

SS infantry were stern opposition to the riflemen. A MG 42 in the medium role mounted on a Lafette tripod.

fields waiting and waiting. As every shell came over and landed amongst our vehicles it was inevitable that we became more and more edgy and the odds against being hit getting shorter and shorter. I was sitting in front of my carrier … chain-smoking, drinking tea, chewing this and that, doing anything to try and keep calm and not showing the nagging fear that we and indeed all of us, were feeling, when a shell landed no more than a few feet away. The carrier rocked heavily as the shrapnel smacked against the armour plating and we were covered with earth and dust, ruining our cups of 'char' and frightening the life out of us. We recovered after a few minutes, cracked a few stupid jokes and said silly things but somehow I felt the tension had lifted from us, as if that was the shell that was meant for us, but it missed, and the odds thereby lengthened quite considerably against us receiving a direct hit. It was a strange feeling and an encouraging one for me.

Before the morning was over, I went to the 02 half-track [Coy 2iC's] to glean any news, when an 88mm anti-tank shell went straight through the vehicle with a hell of a bang. Fearing the worst, I dashed to open the door and found the shell had passed right through the radio transmitter set. The operator, Rifleman Hampshire, had miraculously bent down to pick up a pencil that had dropped on the floor a fraction of a second before the shell hit the truck and although he had a few superficial shrapnel wounds in his back he was okay. The shell would have otherwise gone straight through him.[2]

As a result of an overnight patrol mounted by 8 RB, and confirmed by observation at daylight, the enemy were known to be in strength in Four and seen digging in around Bras with armoured support, which was confirmed when the Cromwells of C Squadron, 2 Northamptonshire Yeomanry, came under fire. This was the *Leibstandarte*'s SS infantry and several surviving *Sturmgeschütz* from Bras and Hubert-Folie.

The previous day, *Sturman* Jospueit, a MG 42 gunner in *Obersturmführer* Kübler's No. 7 Company, II Battalion, 2 SS *Panzergrenadier* Regiment, had been in the woods around Sequeville, 2 miles south of the Bourguébus Ridge, and had endured the bombing at the start of GOODWOOD. Such were the losses that the company was reorganised into two rifle platoons:

> We began our move across the open terrain. It turned into a horror. Leaving plenty of distance between the sections, our column of riflemen moved along the end of a forest. We made our way past the bodies of our fallen comrades. From behind we would hear, 'That's Schutze So-and-So. Dead' … A fighter-bomber roared toward us and headed straight for me. I ran for my life. I thought it was on fire, and I didn't want it to crash on me. The plane, however, suddenly climbed back up in the air. Only then did I realize that it hadn't been on fire; it had been firing with all its guns. I had seen the pilot's face, he had come so close to me.[3]

Arriving on the ridge in the afternoon of 18 July:

> We lay down on the ground and dug it out, sometimes using our bare hands. Then we heard the droning of the planes again. We crawled into the ground. We felt it getting hot around our backsides. The exploding bombs rained kernels of barley from the adjoining field down on us. Once again, we had just barely escaped.
>
> Our only goal was to get deeper into the ground. We dug for our lives. Then, suddenly, our Panthers rolled over our positions toward the front. The *Kommandants* were only looking out for fighter-bombers.

Bras was occupied after dark, as recounted in the *Leibstandarte*'s history:

> At nearly 20.00 hours, the III/1 [SS *Panzergrenadier* Regiment] arrived. It received orders from the Divisional Commander in person. It was to capture Bras and hold it. From there it was to establish contact with the *Sturmgeschütz abteilung*. The situation was unknown in that direction, and we could clearly hear the sound of fighting. As a result, the *Bataillon* commander, *Hauptsturmführer* Graetz, moved the *Bataillon* in combat formation across the line Hubert-Folie–Troteval towards Bras. In the *Bataillon*, the 9. *Kompanie* moved forward on the right, the 11. *Kompanie* on the left, and the 10. *Kompanie* in the middle. About 23.00 hours, they reached the village and found abandoned and fortified positions on its northern edge. Extremely heavy artillery fire began to rain down on our forces at that point.

The cuff title and badge of the *Leibstandarte* that were originally Hitler's personal bodyguard but had for some time been an elite Waffen SS panzer division.

SS-*Brigadeführer* Wisch, commander of the 1st SS *Leibstandarte* Panzer Division, the arrival of which blocked VIII Corps' advance across the Bourguébus Ridge.

The I/1 SS *Panzergrenadier* Regiment held Soliers with a company in Hubert-Folie and the whole defence being backed up by 1 SS *Panzer* Regiment located behind the crest of the ridge. Overall, VIII Corps noted in their history that: 'It appeared that the enemy was not offensively inclined, but that he intended to cling on firmly in his preprepared positions and resist any further expansion southwards.'

Sergeant Hicks provides an example of a well-constructed German bunker while waiting to attack Bras:

> I got out of the carrier and went into the hedgerow to see what could be seen of the countryside beyond leading up to Bras. It was mostly very open large fields criss-crossed with hedges and a smattering of trees. The hedgerow I was standing in was quite substantial with thick growth about 4 feet wide and as I glanced along it, I noticed a bit of a gap and on closer examination I found the entrance to a dugout with a makeshift ladder going down it. I got a torch from the carrier and shone it down the hole and as there were no signs of life in there I climbed down to have a good look at it.
>
> To my surprise it was about 10 feet deep and opened out into a sizeable room with two recesses cut out in the side for sleeping in. It obviously had been vacated in a hurry as there were pieces of equipment, clothing and personal items, photos and scented cards from Paris lying on a makeshift table.

Plans and Preparations

With the three armoured divisions coming up into a line below the Bourguébus Ridge and the flanking infantry formations pushing forward to right and left, VIII Corps HQ wrote:

> The Corps intention for 19th July was in no wise changed, for it was clearly impossible to stay in the awkward and exposed salient so far achieved. At all costs the commanding ground before which 8 Corps was spread as in a diorama must be occupied … The plan remained, therefore, for the armoured divisions to push on to their objectives, whilst 2 Canadian Corps opened the routes from Caen through the suburbs on the east bank of the Orne, and I Corps continued operations against Troarn.[4]

Following a warning order issued at 1030 hours, at 1200 hours the corps commander gave his orders at the tactical headquarters of the 11th Armoured Division. His revised plan was:

(a) At 1600 hours 11 Armoured Division would undertake the difficult task of reducing Bras on its eminence, despite the line of 88mm guns backing it, and thereafter its satellite village, Hubert-Folie.

(b) Guards Armoured Division was to attack Le Poirier at 1700 hours, and exploit towards Frénouville but not to attempt further operations towards Vimont.

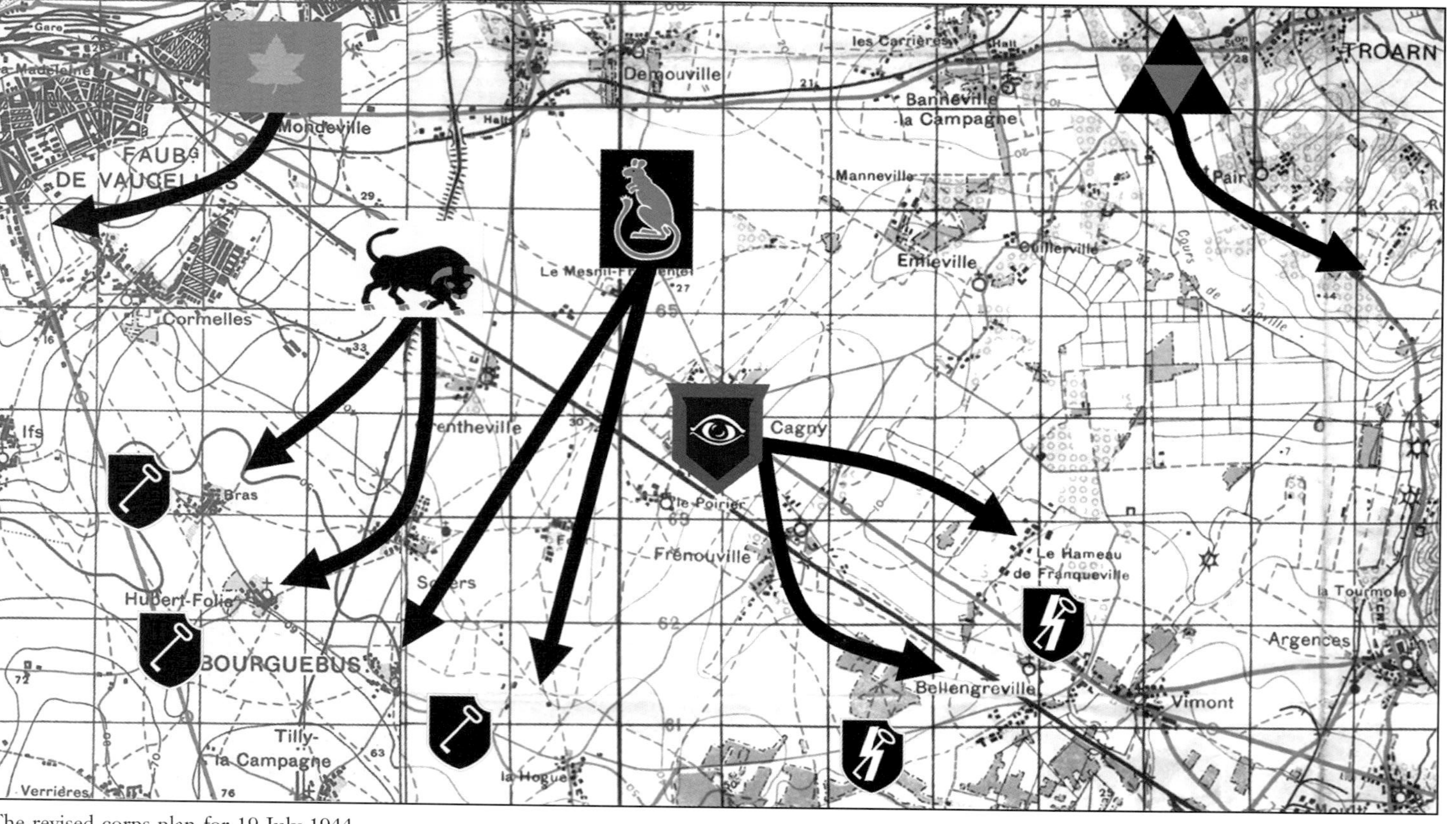

The revised corps plan for 19 July 1944.

(c) At 1700 hours the comparatively fresh 7 Armoured Division had the major role of completing the capture of Soliers and then advancing against Four and Bourguébus. Subsequently, if conditions were favourable, exploitation towards Verrières would be attempted.

All these attacks were to have full artillery support.

Following the previous days' fighting, the 11th Armoured Division's objectives were strictly limited to the capture of the villages of Bras and Hubert-Folie on the Bourguébus Ridge by battlegroups of 29 Armoured Brigade, with battalions of 159 Bridge coming forward to take over their defence when captured. The tanks were to lead the advance and create a break-in into the villages, at which juncture the riflemen of 8 RB would clear the buildings of the enemy defenders in detail.

The three motor companies reverted to the command of Colonel Hunter for the capture of the villages. By 1500 hours the battalion had concentrated just west of the iron ore railway. The CO climbed the embankment to get a view of the two villages and the approaches to them, before summoning the company commanders to the vantage point to give his orders. The plan for the advance towards and the clearance of Bras was for H Company to deploy on the right and F Company on the left. E and F companies would initially be in reserve before clearing Hubert-Folie in the second phase.

The Attack on Bras

For the attack, 151 (Ayrshire Yeomanry) Field Regiment joined 13 RHA in providing artillery support for the assault on the villages. Their war diary provides a useful summary of the fires that helped the riflemen and the tanks:

> Attack went well – CRA ordered smokescreen at 1605. This was stopped by Captain Rolfe as it was obscuring BRAS itself. 2 N Yeo then ordered to attack HUBERT FOLIE approx 1820 hrs. Captain Rolfe ordered U Tgt [UNCLE Target] and smokescreen in area S of IFS to protect right flank of 2 FF and 2 N Yeo from A.Tk guns. This object successfully achieved.[5]

At 1600 hours, under cover of the yeomanry gunners' smoke, the battalion advanced mounted in its half-tracks around the right flank, having difficulty in negotiating the Caen–Vimont railway. However, by 1625 hours the assault companies for first phase had joined the remnants of the 2nd Northamptonshire Yeomanry on the right and began to advance on the village. They were followed by E and G companies, in company with 3 RTR.

The 2 N Yeo's historian described the advance on Bras that began somewhat later than planned:

> The Regiment set off with C Squadron Right, A Squadron Left, B Squadron in reserve. The main road from Caen to Falaise was the Right boundary and the Canadians were reported to be approaching Ifs in the valley beyond it. From the Start Line to the Objective was only about 1,000 yards, but across completely open ground covered with high crops to the ridge beyond. Heavy

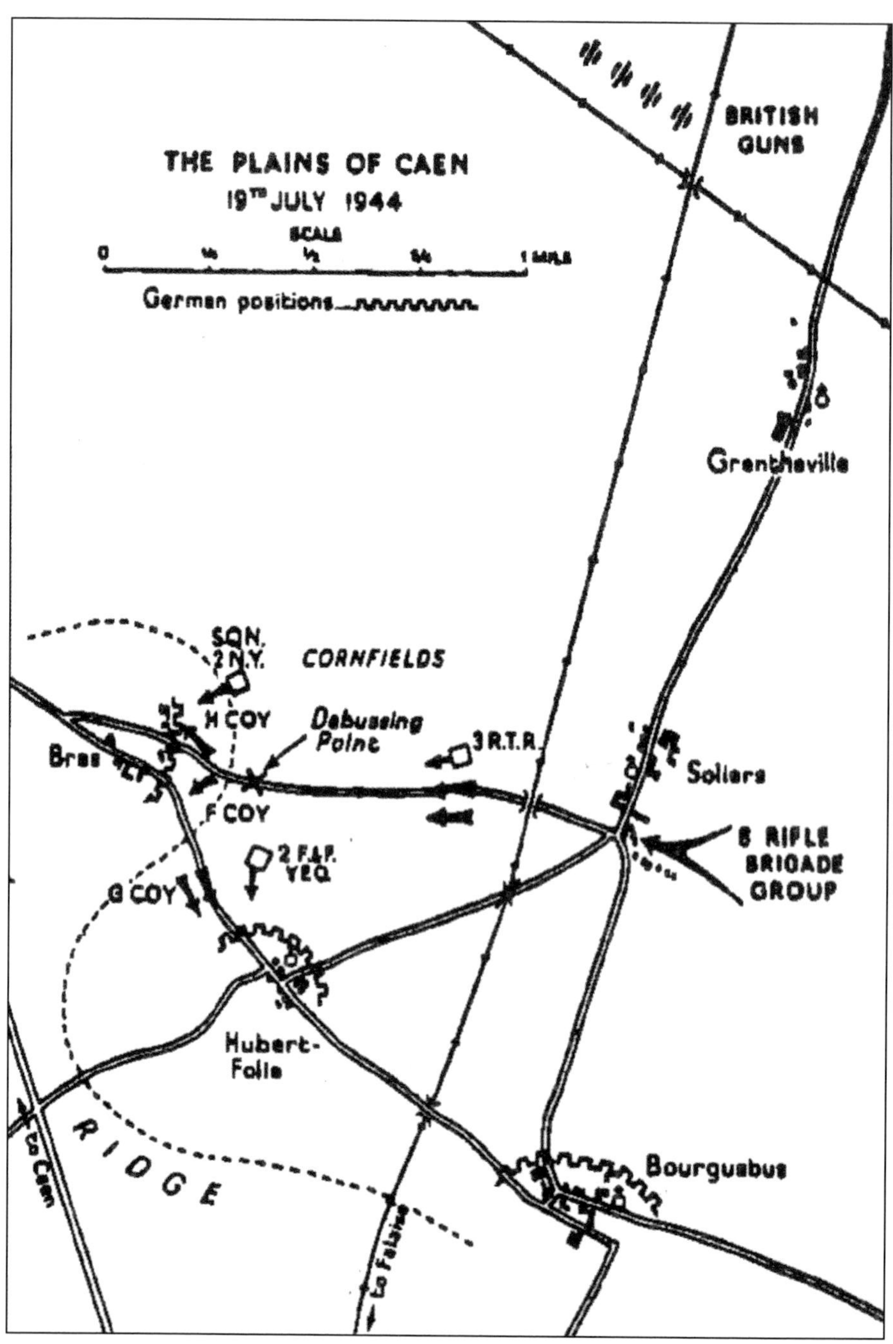

The regimental history's map of the attacks on Bras and Hubert-Folie during the afternoon of 19 July.

mortar fire greeted the advance, a smoke screen put down on the right flank blew back making visibility very poor; while the crops were studded with foxholes filled with snipers and bazookas [*panzerfausts*]. The enemy fired until in danger of being run down and then readily gave themselves up.[6]

The advance of 2 N Yeo, however, faltered before it reached the village due to anti-tank fire from elsewhere on the ridge that was knocking out the Cromwells in the open. Colonel Silvertop saw that 2 N Yeo were held up and offered to help by redirecting 3 RTR's attack onto Bras from the east. The route they would take was both shorter and less exposed to anti-tank fire. This proposal was agreed by Brigadier Harvey and at 1620 hours the Shermans advanced on Bras, followed by the two motor companies astride the Soliers–Bras Road. In his post-action report, *Obersturmführer* Zelinka of 9 Company described his view of the attack by the two tank regiments:

> There was heavy preparatory fire, then the enemy broke through across the main route between Falaise and Caen. The point of breakthrough lay in the sector held by the unit adjacent to us to the left. From this position, the enemy [2 N Yeo] attempted to attack our *Bataillon* in the flank and rear. Hidden from the enemy by the tall grain, most of the *Kompanie* to our left managed to pull back to the positions held by the *Reservekompanie*. Most of our *Kompanie*, however, lay on completely exposed terrain. With a huge number of enemy tanks moving toward us, most of the *Kompanie* stayed put. Additional enemy tanks arrived [3 RTR] and succeeded in penetrating the village of Bras from both flanks. The enemy infantry soon followed. With that, it became impossible for us to pull back to a position level with the *Bataillon*. Only a few individuals made it back to the *Bataillon*.

A film frame showing one of 8 RB's carriers amidst standing crops of the GOODWOOD battlefield.

The 3-inch Mortar

Often referred to as the infantry 'commanding officer's own artillery', in common with other battalions, 8 RB had six 3-inch mortar tubes but in the case of motor battalions they were permanently deployed in detachments of two tubes each to the motor companies rather than often being grouped as a platoon.

The 3-inch mortar was a very simple weapon consisting of four main components that were normally mobile in a Universal Carrier but could be man-packed when necessary: the barrel or tube, bipod, baseplate and sight. It fired a 10lb bomb, its range being determined by the addition of propellant charges. The Motor Battalion pamphlet details the 3-inch mortar's characteristics:

(a) The mortar, at present, has a maximum range of 2,700 yards.

(b) It has a very high trajectory and is therefore affected by wind.

(c) The burst of the bomb is effective for 100 yards all-round the point of impact.

(d) It is capable of rapid changes of direction within an arc of about 36 degrees without moving the bipod.

(e) It is easy to conceal and is capable of being fired from behind high cover.

(f) It can be fired at night or when blinded by fog or smoke.

(g) Once ranged, it is capable of sustained and very accurate fire without overheating. Normal rates of fire are rapid, 20 rounds per minute: slow, 6–7 rounds per minute.

(h) The flash is negligible.[1]

In a motor battalion the detachments were normally commanded by a subaltern officer who acted as the fire controller forward with the company commander or in a position of good observation. Target information was passed back to the mortar line by voice, radio or if possible defence telephone. The mortars were normally commanded by a sergeant. When the battalion was undermanned along with the second officer in the scout platoons, the mortar detachments were the first to be without an officer, as was the case for 8 RB for much of the Normandy Campaign. The Motor Battalion pamphlet describes the role of the 3-inch mortar detachment:

(a) In conjunction with the artillery plan, mortars should be used to
  (i) Neutralise areas held, or thought to be held, by the enemy, either with HE or smoke; these may be on a flank.
  (ii) Assist in the firefight by giving covering fire on located enemy localities.
  (iii) Move forward behind the assaulting troops and assist in consolidation of the objective.
  It should be remembered that the burst of a mortar bomb is outwards and upwards; only a direct hit will have any effect on emplacements, and a direct hit cannot be expected without the expenditure of a considerable amount of ammunition. However, since fragments of the bomb radiate in all directions very close to the ground there is a good chance of causing casualties to troops who are dug in and manning their weapons with their heads above ground.

A 3-inch mortar crew and carrier photographed for a training publication.

(b) When opposition is encountered the mortar detachment commander, if not already with him, should join the commander of the force on his reconnaissance and advise on tasks. The detachment should meanwhile be moving up to a rendezvous.

Sergeant Hicks of H Company's detachment described how they made life working from their Universal Carrier more comfortable:

As our Bren gun carriers were open-topped tracked vehicles we were very vulnerable to the vagaries of the weather, whereas all the other platoon vehicles were American half-tracks with very good weatherproof canvas-type

The mortar barrel and bipod were carried on the rear of the carrier and baseplate on the front, with ammunition racks taking up much of the rear crew compartments.

hoods. It was not long therefore before we had 'organised' some excellent tarpaulins which were strapped to metal poles along the side of the carrier and which could be extended out to form very effective lean-to type tents. They were particularly useful when we had to dig and sleep in slit trenches or when preparing food on our portable cookers in the pouring rain and they played no small part in making our lives that much more bearable when circumstances allowed. They survived right through to the end of the campaign though very much patched up to cope with all the shrapnel and bullet holes and tended to leak quite a bit in heavy rain.

The leading tanks of the 3 RTR squadrons were reported to have broken into the village at 1646 hours but they were being engaged by the SS infantry armed with *panzerfausts* and the other squadron to the east of the village was being engaged by anti-tank guns from Hubert-Folie. Understandably, Colonel Silvertop wanted the support of the 8 RB companies, but they were not on his regimental radio net, so this took time. Eventually, with regrouping procedures complete, the motor platoons in their half-tracks advanced through the corn to the village, with the riflemen throwing grenades into the deeply dug German trenches as they passed. The carriers were 'thrashing through the corn like destroyers, rounded up many prisoners on the way in'. The motor platoon's half-tracks halted about 100 yards from the village and the riflemen debussed and deployed for the attack. The 8 RB war diary recorded that:

> BRAS turned out to be very strongly held by approx. one Bn well dug in, but 3 RTR did magnificent work and, coming down from the NORTH, were able to report … by 1717 hrs that they were in to it, and by 1730 hrs that they had gone right through it. The two Coys were right behind them and immediately debussed and pressed on into the village, H Coy on the right and F on the left with the track running NE from BRAS as the dividing line.

Thanks to the bombardment, rubble was a problem for the tanks in Bras but, working with the infantry, they knocked down walls and blasted entry points for the riflemen in buildings with, for their size, their remarkably effective 75mm HE rounds.

> The enemy belonged to 1 SS Pz Div … and they fought bravely as they were expected to fight. However, owing to the speed with which we got into the village they had little time to collect themselves. Some came up out of their slit trenches with grenades in their hands, and were soon dealt with; inside the village there were still MG posts and snipers holding out, and they too were speedily mopped up. F Coy pressed right through the village and were in time to get a good shoot at a party of about 70 who were escaping SW but for the rest it seems that very few of the enemy got away. The speed of the attack and the mopping up was such (the village was reported clear by

Riflemen of 8 RB clearing a village in Normandy.

1800 hrs.) that almost the garrison was killed or taken prisoner and the 3rd Bn 1 SS PGR was considered largely written off.

Citation for awards to riflemen add more detail. Corporal Fitzgerald MM was in action again with 6 Platoon of F Company:

Later on the same day Cpl. Fitzgerald's section was the first one into the village during the attack on Bras. Showing great dash and initiative, he cleared his sector of the village with speed, taking the enemy by surprise and capturing twenty-five prisoners.

He was a fine example and inspiration to all those with whom he came in contact.

Lieutenant Sedgewick followed with the rest of 6 Platoon, earning the Military Cross:

On 19th July F Company was the leading company in the attack on the village of Bras. Lieut. Sedgwick's platoon, now considerably reduced in numbers, was again in the lead. Although the enemy was in strength in the village, Lieut. Sedgwick led his platoon with such skill and speed that a considerable number of the enemy were killed and the remainder taken prisoner. He showed great dash and determined leadership in overrunning two orchards full of the enemy and capturing twenty prisoners and twelve *nebelwerfers*. Throughout the evening and the ensuing night Lieut. Sedgwick's platoon were under heavy artillery and mortar fire and he showed great coolness and courage, being a really good example to his men.[7]

In the right half of the village, H Company were in action. Again, a citation for the MM provides some detail:

> On 19th July L/Sergt Triggs took part in the attack on Bras. With great dash he led his carriers into the village and captured an anti-tank gun complete with crew. This anti-tank gun had been causing casualties to our armour [2 N Yeo. L/Sergt Triggs then took up position with his section covering the exits from the village].
>
> During both days' fighting L./Sergt. Triggs showed great coolness under heavy fire and shell fire and was an inspiration to his section and to all around him.

Gunner Prett was the only unwounded member of the crew of an OP Sherman belonging to 151 Field Regiment that had been knocked out during the bombardment of Bras before H Hour and had been taken prisoner by the SS. The regiment's war diary describes how he helped and was helped by the riflemen of H Company:

> Gnr Prett and the wounded men were then taken prisoner and taken to BRAS, where they were put in a burning house under escort. As the tanks of 3 RTR entered the village the [PW] escort disappeared but the exit to the house was covered by snipers. Gnr Prett, having dressed their wounds, restrained the wounded men from leaving and finally attracted the attention of a coy of the 8 RB to whom he gave information about the snipers. The wounded were then evacuated by 8 RB. By his coolness and quick thinking Gnr. Prett undoubtedly saved the lives of his companions and gave valuable assistance to the 8 RB in pointing out the snipers.

Bras was fully cleared by 1800 hours, with F Company, as already mentioned, getting a good shoot at the withdrawing enemy from the southern extremity of the village. However, while waiting for the attack on Hubert-Folie, they had 'a very unpleasant time under enemy shell and mortar fire' and ten minutes of fire from four medium regiments that should have been firing on Hubert-Folie!

Moving up to Bras, Sergeant Hicks had to first clear the area before he could deploy his detachment's mortar line to support the attack on Hubert-Folie:

> I moved our two mortar carriers up to the edge of the village and stood up to get a clearer view of the surroundings to find the best position for our mortars when a sniper bullet smacked into the armour plating only an inch below the top edge. Boy! Did my arse hit my seat in record time and it was some minutes before I ventured to put my head up over the top and eventually get out, still inwardly shaking with fright.
>
> The Germans were very good in preparing defensive positions and concealing them with appropriate camouflage. I saw my old 16 Platoon

commander Lt. Jeff Coryton, about 30 yards away talking to Sgt. Rodwell so I went over to have a quick chat and glean what news I could from him and someone shouted a warning. Part of the cover of a slit trench which had been completely concealed had moved and a hand with a Luger pistol in it was aiming towards Jeff only a few feet away. Fortunately Jeff beat him to it and shot him in the head. Poor Jeff was visibly shaken and truly upset by the fact that he had just killed his first German and he only just managed not to puke his heart up.

We investigated the slit trench further and another German scrambled out with a grenade partly concealed in his hand but Sgt. Rodwell spotted this, tripped him up and shot him twice through the chest. Must admit all this close quarter bloodshed was making me turn a whiter shade than pale.

To obviate any possible further trouble from this concealed trench, which was obviously quite deep and large underground, I tossed two hand grenades into it and we had no more bother from it nor did we trouble to investigate it further.

Staff officers of VIII Corps examine a captured *Leibstandarte* senior NCO's jacket.

Hubert-Folie

With 3 RTR having delivered the attack on Bras instead of 2 N Yeo, the plan was quickly adapted. At 1703 hours, the already depleted F&F Yeo, with just twenty-five tanks, were placed at immediate notice to move from reserve and advance on Hubert-Folie from the east with the riflemen of G Company, while 2 N Yeo were ordered to reorganise and be prepared to support the attack on Hubert-Folie. It took some time for orders to be given and for 3 Monmouths from 159 Brigade to come up and take over the defence of Bras 'once mopping up was complete'. In the meantime, Hubert-Folie was pounded by the divisional artillery and the 5.5- and 4.5-inch guns of the AGRA that had been redirected onto the correct target. The battalion's own 3- and 2-inch mortars fired smoke to cover the infantry assault.

The N Yeo disengaged from Bras and redeployed mustering about a single squadron of tanks:

> A Squadron was sent to the South of Bras to support them into the village and B Squadron sent a troop to watch the Right Flank beyond the main road. This all took some time and was in full view of the enemy, who had withdrawn their armour to a higher ridge to the South. As a result there was continuous observed fire by mortars and medium artillery, and long-range AP shells kept on knocking out tanks in spite of their continually changing position.

Meanwhile, 8 RB were waiting in Bras, as recorded in their war diary:

> As soon as the mopping up was complete, 3 Mons came up and took over the village, and we turned our attention to HUBERT'S FOLIE [*sic*]. Unfortunately, there was a delay in the execution of the attack and we found ourselves waiting about at BRAS for two hours being shelled intermittently by enemy mortars ... They were held up for a while by what seemed to be Bren fire from a Sherman.

The F&F Yeo also recorded this incident in their war diary. 'One of the A/Tk guns, previously reported destroyed, suddenly came to life again and brewed-up 3 tanks at point-blank range, before being finally and irrevocably liquidated.'

It was now G Company's turn to advance across the 500 yards between Bras and Hubert-Folie mounted in their half-tracks, covered by the 25-pounders of the divisional artillery and fire from squadrons of the two yeomanry regiments. The regimental history provides some detail:

> A quick plan was made for the high ground south of Bras to be smoked off and for supporting fire to be brought down by two field regiments and a medium regiment on to the orchards on the north edge of Hubert Folie, subsequently lifting on to the centre of the village. The attack was to go in from the north-west of Bras, directed almost due south on to Hubert Folie. It started at eight o'clock. The leading squadron arrived on the edge of the

What is left of the open ground between Bras and Hubert-Folie. G Company attacked to the right of the road where the houses now stand, and two squadrons of F&F Yeo were in the corn to the left.

village before the barrage had finished: G Company were right behind the tanks and out of their half-tracks like lightning.[8]

Major Bell recalled that:

A very effective smoke-screen was put down by our 2-inch mortars and behind this 10 and 11 platoons formed up and went into the attack, followed by 12 Platoon, who were to consolidate in the rear. The village had been previously shelled and the tanks were pumping stuff into it too. As the motor platoons moved in. I called over the air for the tanks to stop firing, but one went on firing and nothing could be done to stop it. It later transpired that this was a tank knocked out the day before and a German was manning its machine-gun. The carriers who were acting as flank protection ahead of the motor platoons came under fire from this machine-gun, and Cpl Isard, a very old and popular member of the Company, was killed. We later had the satisfaction of seeing this German 'brewed up' in no uncertain manner at very close range.

Meanwhile, the clearing of the village was going well, and we were soon able to announce that it was in our hands completely and that we had collected a fair haul of prisoners as well as killing other enemy.

During this operation, my command vehicle was standing on the village green and, thinking that morale amongst my crew had not yet quite recovered from the previous day, I told my rear gunner, L/Cpl Hodgson, that I had seen movement in a clump of bushes close at hand. Whereupon his Sten gun blazed into action and aggression became the keynote again. I have never dared to tell him that it was a put-up job, and I hope if he reads this, he will not feel badly about it. At any rate, it produced exactly the effect which I had hoped it would.

The war diary concluded: 'The clearing of the village took a short time; it appeared to contain only a few stragglers from BRAS, one of whom was the Bn MO of the SS Bn who proved extremely useful to the Inf Bn who took over from us [4 KSLI].' In all about eighty prisoners were added to the battalion's bag. Finally, 'By 2115 hrs the battle of BRAS and HUBERT'S FOLIE was over.'

At 2315 hours, G Company had handed over the defence of Hubert-Folie, 'enabling us to return once again to the railway embankment, where we rejoined the rest of the battalion in laager', but this was not before the Luftwaffe made another appearance at 2130 hours, just before last light:

We had had a lucky escape when thirty Focke-Wulf 190s appeared overhead, but they evidently had no clue as to what was going on and made off, dropping their long-range petrol tanks amongst us, and our gas experts rushed for their respirators.

A captured *Schwimwagen* receiving some rebadging by soldiers of 23 Hussars. Most such vehicles were taken off units by the military police before crossing the Seine.

The Fw 190s' main target was brigade headquarters, but their ordnance overshot, causing no damage.

Colonel Hunter's post-operational report summed up 8 RB's achievements in Operation GOODWOOD and the cost in lives and material:

> The Bde retired for the night to the bottom of the hill from where we had set out, G Coy waiting behind for half an hour until the Inf Bn. had taken over. The next day we were withdrawn [to the Giberville area] owing to tank casualties and the Bn had a well-earned rest. During the two days of Op GOODWOOD the Bn had taken 581 PW and had captured or destroyed the following items of enemy eqpt: – 27 Nebelwerfers, 5 Hy A/Tk guns, 8 ½-tracks, 1 SP Hy Inf gun and lastly 1 Volkswagen whose insignia of 1 SS. Assault Gun Bn was soon replaced by those of F Coy 8 RB. Our own casualties during the operation were 4 Offrs wounded, seven ORs killed, 75 wounded, and 2 missing, and in vehs 4 ½-tracks and 1 carrier lost through enemy action.

The first day of GOODWOOD had cost the brigade 115 tanks and 'if 19th July had not been so costly as the previous day, nevertheless 65 tanks had been knocked out in the fighting, and of these 37 belonged to 2 Northamptonshire Yeomanry'.[9]

6-Pounder Anti-Tank Gun

The British anti-tank gun at the beginning of the war was the 2-pounder. The German equivalent was the Pak 36/37 3.7 cm anti-tank gun, which had an acceptable performance until 1941, when it proved ineffective against the Soviet T34. This earned it the nickname 'Panzer knocker'. With more heavily armoured tanks being produced there was soon a demand for improved performance, and the results were the British 6-pounder (57mm) and the German 50mm PaK 38 anti-tank guns, which became the norm by 1942. However, armoured developments, particularly with the Red Army T34 tank, spurred on the production of the German 75mm PaK 40, along with better armoured tanks in a race between penetration and protection.

The 6-pounder, as used in motor battalions, was towed by a carrier and had a crew of six, three of which were needed to serve the gun. Unlike the 2-pounder, the new gun was mounted on a conventional two-wheeled split trail carriage on pneumatic tyres, but without a spring suspension.

The effective range was quoted at 1,650 yards, but with normal armour piercing rounds the chance of penetrating the frontal armour of the Panzer IV was minimal. However, the gun could and did knock the heaviest German tanks via the side armour, but at lesser ranges.

In defence, the gun was invariably dug-in to just below the level of the barrel for protection, but the large open pit, in common with other towed anti-tank guns, was particularly vulnerable to airburst shells.

The establishment for 8 RB was six 6-pounder guns in three platoons (17, 18 and 19 platoons).

Work on improving the performance of the ammunition went on in parallel with improvements to the gun barrels and sights. Various projects were worked on, but it was Armour Piercing Discarding Sabot (APDS) which, after protracted development and testing, was finally issued in Normandy, massively improved the 6-pounder's performance, close to that of the German Pak 40. The APDS was, however, limited in supply.

The concept was to use a hard and heavy metal for a penetrator of a sub calibre, which had less drag through the air and delivered a greater impact velocity and therefore improved armour penetration. To bring the round up to full 57mm calibre for firing, the penetrator was surrounded by a sabot consisting of three petals of a lighter metal that fell off having left the barrel, allowing the unimpeded penetrator to speed on it way. APDS rounds improved penetration but at the expense of a marginal loss of accuracy at longer ranges. Despite its excellent performance, there was another problem for troops who had not been trained in its use: when the discarded sabot petals hit the ground short of the target, there was an erroneous impression that the gun was firing low.

Chapter Eight

A Move West

Having been pulled back to the fields between the ruins of Giberville and Démouville, the rain that came down in torrents and brought GOODWOOD to a halt promptly turned fields into a thoroughly unpleasant sea of mud. Major Cunliffe of F Company recalled that:

> At Démouville we found ourselves in plough. I chose the position of Company Headquarters in a dip; I thought wisely out of the wind but subsequently this proved fatal. We had grown used to dust and sand and their inconveniences, and now we had our first taste of mud, real, rich, cloggy mud, and then it started to rain, and rain as hard as it does at Old Trafford. We began to build elaborate bivouacs and erect tarpaulins, but Company Headquarters for once was hopelessly placed, and we were soon swimming in water. It was useless wearing ammo boots, we had no gum boots, and so I walked around in bare feet. The Volkswagen looked pathetically wet and seemed anxious to demonstrate its amphibious powers. Eventually we retired, or should I say were towed, to a more pleasant place.

On 22 July, with a strong enemy defence clearly in place east of the River Orne, 8 RB joined the rest of the division, recrossing the Orne via LONDON Bridges and passing a column of jeeps and motorcycles bearing the Prime Minister and General Montgomery. The battalion occupied paddocks and orchards around a village that had been squarely in the path of the Canadians during Operation CHARNWOOD during 8–9 July:

> The more pleasant place was Cussy, a badly bombed village with a lovely old Abbey [Ardenne], not far from Caen itself. Here the weather was calm, and the routine consisted of sleep, the maintenance of vehicles and weapons, the washing of bodies and clothes, and then more sleep.

Major Bell described life for G Company after catching up with sleep:

> Our week at Cussy was very pleasant and nothing sensational happened. Much work was done and at the same time much rest and relaxation were also able to be included. Liberty trucks were run to Bayeux and we were given our first vacancies at rest camps,[1] which were greatly enjoyed by those fortunate enough to go there. A notable event was the appearance of a N.A.A.F.I. mobile van, the first we had seen since leaving England, and its contents were very quickly sold out.

An ENSA show in the Normandy lodgement after D Day.

A cinema was set up in a barn in Cully and enjoyed by the riflemen despite the many holes in the roof that let light in and made it hard to discern what was happening on the screen.

While the motor companies were at their ease, Battalion Headquarters was, however, constantly planning and replanning as a series of tasks were passed down from divisional headquarters. For instance, on 27 July the brigade was placed on four hours' notice to move to support 4 Armoured Brigade against a 'possible but unlikely counter-attack in the Mouen area', and on another occasion, 8 RB was warned for attachment to 4 Armoured Brigade to strengthen their defence on the Odon.

Despite receiving forty reinforcements on 21 July, 'mainly' as the war diary noted 'from 8 Corps Def Coy',[2] the battalion was still badly under strength.[3] At a conference held at divisional headquarters on 23 July, among the many issues raised as a result of Operation GOODWOOD, was the manning of 8 RB:

(d) Owing to reinforcement difficulties, the Motor Bn. may have to be reduced. (a) Is the A.Tk gun element necessary, (b) are more carriers required?

On account of absence of reinforcements, the Motor Bn is now organised as Sp Coy [E Company] and three Motor Coys, each of a Scout P1 of 8 carriers and two motor pls. This temporary arrangement is considered satisfactory. The A/Tk gun pl in Sp Coy should be maintained. More carriers would be useful but NOT until the necessary specialist reinforcements are available to man them.

It would not be until 10 August, after further significant casualties, that 8 RB received meaningful reinforcement. In the meantime, for F Company, for example:

In spite of the arrival of more reinforcements, the casualties of the last two battles made it necessary for the Company to be reorganised with only two Motor Platoons – these were known as 6 and 8. The Scout Platoon remained at full strength.

The minutes of the post-operational conference also recorded decisions on the division's tactics and included a substantial paragraph on the attack and clearance of villages:

(a) Co-operation and drill for dealing with a village or strong pt with the Motor Bn/Coy:-
It will be normal for a Motor Coy to be under comd of the Armd Rest concerned. This Motor Coy will be netted on the Regt Comd Net. Its carriers should move as close behind the tks as possible and in any event not more than 500yds. The motor pls should move 500yds behind the carriers. Tks will penetrate the outskirts of the village probably before arrival of inf in order to protect themselves from AP fire from the flanks and one or two tps will be prepared to penetrate the village with inf. Close liaison between tp and pl comd. will be essential. Only very limited objectives in the village will be taken on. The remainder of the rest [of the squadron] will take up suitable posns in the outskirts of the village and will be relieved as early as possibly by a troop of RA A/Tk guns, placed under comd and netted to the motor company. RHA smoke will be used on the flank to mask A/Tk guns covering the armd regts' approach to the village and the possibility of using smoke in the village itself is not ruled out. It is a decision to be made by the commander on the spot bearing in mind that there will already be considerable haze and dust from the artillery barrage, and an addition of smoke may complicate the infantry's task considerably. In any event, it might be extremely useful on the far side of the village or for smoking off a part of the village with which the motor company does not propose to deal immediately.

To address some of the procedural issues that caused delay in regrouping at Bras and Hubert-Folie, the location, communications and movement of the company commanders was also discussed and resolved. In order to reduce the size of the

Infantry mounted on tanks to form a battlegroup during Operation BLUECOAT.

armoured regiment's tactical headquarters, which were obvious and attracted a lot of fire during GOODWOOD:

> The Motor Coy Comd will in future move with his Coy and not with the Armd Regt Comd. He will however be on the regimental command net and is therefore quickly available in a scout car. In the event of armd action developing, the Motor Coy [will] move out of the immediate vicinity in order to be clear of the passage of AP shot.

While the tactics recommended were taken on, Major General Roberts had been mulling over the grouping of infantry and armour within the division.

Infantry and Armour Battlegroups

Major General Roberts' own experience of fighting in relatively small areas of bocage during Operation EPSOM and the considered opinion of 7th Armoured Division following the advance south of Bayeux and further west around Villers-Bocage were being discussed by British armoured formations. An 'Immediate Report' of 17 June (post Operation PERCH) was circulated by XXX Corps highlighting the difficulties:

> It is quite clear from even short experience of fighting in this type of country that the tanks require a great deal more infantry with them than is provided by the Motor Battalion. The Motor Battalion is only sufficient to clear a limited number of localities and protect them at night. In this country where pockets of resistance such as snipers and machine gunners lie and allow the

tanks to bypass them, opening up after the tanks have passed them, it is necessary to have working with the tanks sufficient infantry to flush, capture or kill the enemy who adapt these tactics. The armoured division organisation provides enough infantry for this to be done, and I am satisfied that it is only a matter of evolving correct tactics to tie together the infantry and tanks so that both can work together on perhaps slightly new lines of tactics to achieve what we desire. It is most noticeable that tanks on the move through enemy territory now complain frequently that enemy infantry are infiltrating between their various troops and squadrons. In the past they complained of anti-tank guns at a long range, but in this country they are much more nervous of infantry at close range, partly because of the sniping which they will receive when they have bypassed the infantry and partly because of the feeling of nervousness which they got from passing continuously through close country unsearched for concealed short-range anti-tank weapons and infantry. [4]

The first steps in breaking down the divide between the armoured and infantry brigades and balancing forces had already been reported by 21 Army Group as early as 12 June:

> The solution to this which worked well on the front of 7 Arms Div, advancing on two roads (FOLLIOT – BERNIERES BOCAGE and BAYEUX – TILLY SUR SEULLES) was to put an infantry battalion under command of the Armoured Brigadier and an armoured regiment under command of the Infantry Brigadier, each being responsible for one road.[5]

In the 11th Armoured Division during the wait before EPSOM, 29 Armoured Brigade ordered 'Regts to liaise with bns 159 Bde in view of possibility of close country fighting.' Five weeks later, when faced with a redeployment from the open plains south-east of Caen west to the true bocage, addressing the problems highlighted by XXX Corps, Roberts took the regrouping of infantry and tanks a stage further. The divisional war diary recorded that:

> For the advance which was to start at 0700 hrs on the next day fresh brigade groupings were adopted. The experience of 7 Armoured Division and the Americans in the bocage country had demonstrated the necessity for the closest co-operation of tanks and infantry. In this region of thick woods and narrow roads winding between impassable hedges and ditches, a number of local engagements were anticipated and under these conditions reasonable progress could be assured by the infantry moving with the tanks on all routes and often actually riding on them.[6]

The divisional historian continued:

> We now attempted a closer cooperation than had yet been tried within an armoured division. The division, therefore, moved in Brigade Groups: 29th Armoured Brigade Group on the left consisting of two of their

The main unit cap badges of the 29 Armoured Brigade.

armoured regiments (23 H and 3 R Tks), the motor battalion (8 RB), and one infantry battalion borrowed from 159 (3 Mon); and 159 Infantry Brigade Group on the right with their remaining two units (4 KSLI and 1 Hereford), one armoured regiment (2 FF Yeo) and the armoured reconnaissance regiment (2 N Yeo), now as usual employed as a normal tank unit. For the reconnaissance and contact which our role demanded we were again allotted an armoured car regiment; this time it was the Second Household Cavalry.[7]

For 8 RB in Operation BLUECOAT, this meant that rather than a company being allocated to each of the brigade's armoured regiments, the whole battalion would initially work with 23 Hussars.

The Fourteen-Man Compo Ration

As the campaign progressed, fresh food gradually became available but only in small quantities, with diaries and accounts all enthusiastically reporting the arrival of the first white bread, then English beer and finally meat. The tinned Fourteen-Man Ration or 'Composite Ration', more often known as (Compo), however, remained the staple fare of the riflemen.

Living and working from half-tracks, 8 RB did not have to resort to the 24-Hour Pack issued as landing rations for the first three days in Normandy but throughout the campaign mainly received their food in wooden boxes issued by their company quartermaster sergeants. There were six types of Compo, menus A–G, each of which were to be rotated but in practice it was not uncommon for the same types to be issued for several days in a row, if not longer. More than a few riflemen developed lifelong aversions to some of the compo delicacies, typically bully beef and chicken supreme. Each day's tins and packets consisted of breakfast, a main meal, a snack, plus chocolate and sweets, hard tack biscuits AB (Alternative to Bread), tea, sugar and milk mixture, cigarettes and latrine paper.

Cooking for the riflemen was on a modest petrol cooker, the No. 2 Burner. Cans were punctured and heated in boiling water, which could be used for tea or the powdered soup contained in some menus, but there were many ways in which the food could be prepared. These included negotiating with tank crews to heat cans on their engines and exhaust system.

There was a ban on purchasing food from French sources to avoid creating a market that sucked essential commodities from towns, but in practice a vigorous trade developed for dairy products, eggs, cider and Calvados (apple brandy)[8] in exchange for cigarettes, chocolate and sweets and other less popular compo items.

A No. 2 Burner was a part of the equipment issued with every half-track along with a pan for boiling the cans.

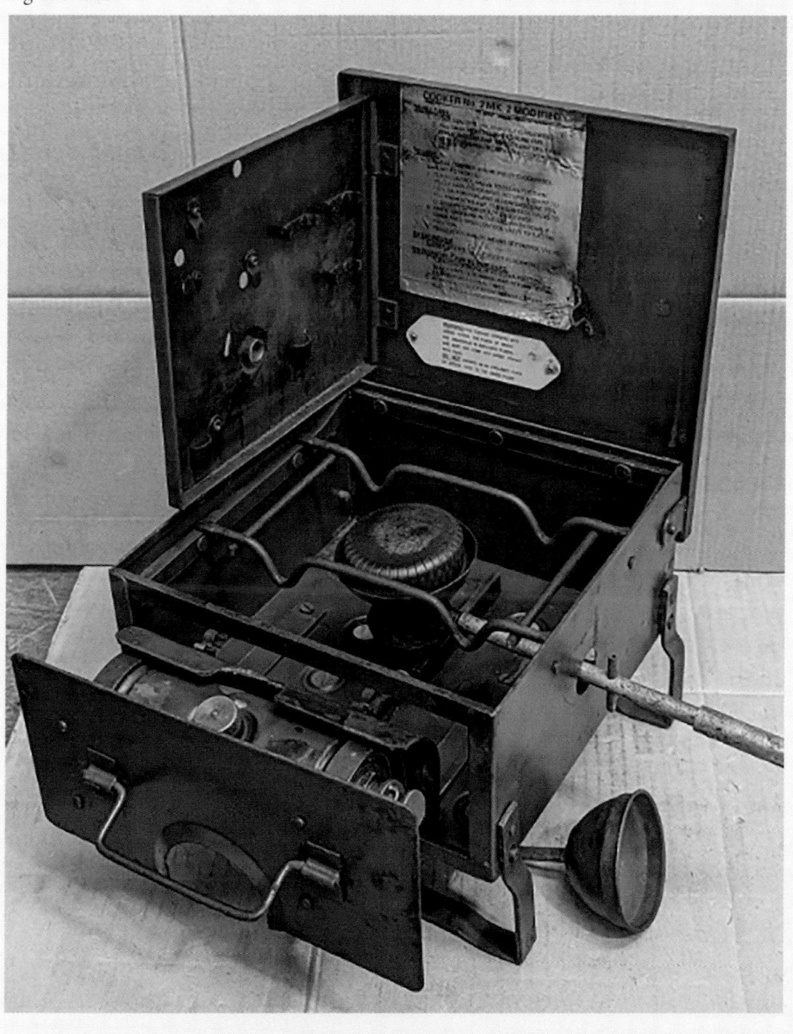

COMPOSITE RATION PACK

TYPE A

(14 men for one day)

Contents and Suggested Use

BREAKFAST Tea   *   3 tins (2 tall, 1 flat—Tea, Sugar & Milk Mixture)
†Sausage (1 hr.)   2 tins
Biscuit   *   1 tin
Margarine   *   1 tin

(* Items marked thus are also to provide for other meals)

DINNER †Steak & Kidney Pudding ($\frac{1}{2}$ hr.)   11 tins
†Vegetables ($\frac{3}{4}$ hr.)   4 tins (2 large, 2 small)
Tinned Fruits   2 tins

TEA Tea   —   (* see above)
Biscuit   —   (* see above)
Margarine   —   (* see above)
Jam   1 tin

SUPPER †Baked Beans ($\frac{3}{4}$ hr.)   3 tins
Biscuit   —   (* see above)

EXTRAS Cigarettes   2 tins (1 round, 1 flat— 7 cigarettes per man)
Sweets   1 tin
Salt   —   } packed with
Matches   —   } sweets above
Chocolate   —   (1 slab per man— packed with biscuit)
Latrine Paper

DIRECTIONS

Tea, Sugar and Milk Powder.—Use a dry spoon and sprinkle powder on heated water and bring to the boil, stirring well. 3 heaped teaspoonfuls to 1 pint of water.

† May be eaten hot or cold. To heat, place unopened tins in boiling water for minimum period as indicated. Sausage may be fried (using margarine) if preferred.

Wt.   /G.2618   1,568M in 7 Sorts   6/42   KJL/652   Gp. 698/3
Wt. 41076/5747   200,625/5 Sorts   12/42   KJL/3222   Gp. 698/3   **J.6393**

The day's menu for two of the types of Compo.

COMPOSITE RATION PACK

TYPE B

(14 men for one day)

Contents and Suggested Use

BREAKFAST Tea * 3 tins (2 tall, 1 flat—Tea, Sugar & Milk Mixture)
†Bacon (1 hr.) 3 tins
Biscuit * 1 tin
Margarine * 1 tin

(* Items marked thus are also to provide for other meals)

DINNER †Steak and Kidney (½ hr.) 10 tins
†Vegetables (¾ hr.) 4 tins (2 large, 2 small)
†Pudding (1 hr.) 3 tins (2 large, 1 small)

TEA Tea — (* see above)
Biscuit — (* see above)
Margarine — (* see above)
Jam 1 tin

SUPPER †Soup, Scotch Broth (1 hr.) 2 tins (mix with equal quantity of water)
— Biscuit — (* see above)

EXTRAS Cigarettes 2 tins (1 round, 1 flat— 7 cigarettes per man)
Sweets 1 tin
Salt — { packed with
Matches — { sweets above
Chocolate { 1 slab per man packed with biscuit
Latrine Paper

DIRECTIONS

Tea, Sugar and Milk Powder.—Use a dry spoon and sprinkle powder on heated water and bring to the boil, stirring well. 3 heaped teaspoonfuls to 1 pint of water.

†May be eaten hot or cold. To heat, place unopened tins in boiling water for minimum period as indicated. Bacon, and pudding cut into ½-inch slices, may be fried (using margarine) if preferred.

Wt.   /G.2618   1,568M in 7 Sorts   6/42   KJL/652   Gp. 698/3
Wt.   /G.2341   1001M in 7 Sorts   9/42   KJL/1987   Gp. 698/3

A Fourteen-Man Composite (Compo) Ration box of tinned and packeted food including cigarettes and toilet paper.

The Break-Out Begins

While 8 RB was out of the line resting and refitting at Cussy, after making slow progress through the thick bocage country to the west of the invasion area, the First US Army successfully launched Operation COBRA on 25 July. Two days later, with First Canadian Army taking over the Caen flank, in his M 515 Directive, General Montgomery gave General Dempsey orders:

> 10. The Second Army will regroup and will deliver a strong offensive on its right wing with no less than six divisions.
> The offensive will be delivered from the general area of CAUMONT. The initial objective will be the area of ST. MARTIN BESACES – LE BENY BOCAGE – FORET L'EVEQUE and a strong force will be held for quick exploitation towards VIERE.
> 11. The sooner this offensive can be launched the better. The latest date, consistent with good weather, will be 2nd August.

However, with COBRA rapidly gaining momentum against weak opposition, General Dempsey was ordered to launch Operation BLUECOAT without delay. For this offensive he would use VIII Corps and XXX Corps, the latter holding the main effort, with the 7th Armoured Division under command. With regard to VIII Corps' plan, General Roberts recalled:

> I received verbal instructions from the corps commander that initially 15th (Scottish) Division would be carrying out the main attack, supported by 6th Guards Tank Brigade (Churchill tanks), with their main objective being

an important feature, Point 309, some 2,000 yards east of a village called St Martin-des-Besaces. We, 11th Armoured Division, were to fight forward, protecting the right flank of 15th (Scottish) Division and in touch with the Americans on our right flank.

As 2 N Yeo had relinquished their recce role and become an ordinary tank regiment under command of 159 Brigade, the division, as noted above, was allocated the armoured cars of 2 Household Cavalry Regiment in their place. The Guards Armoured Division were in corps reserve for exploitation.

Redeployment

At 1510 hours on 28 July, Headquarters 29 Armoured Brigade received a warning order that they were to move west that night. Dispatch riders were sent to gather men of the division from rest centres and the beaches and order an 'immediate return to their units', where after six days' rest, vehicles were being packed.

The 15th Scottish Division, which the 11th Armoured Division would join under command of VIII Corps, was already in place having taken over the section of front immediately south of Caumont from the V US Corps. Likewise, 50th Northumbrian Division (XXX Corps) was holding the front from Hottot west. That left an infantry division, one armoured brigade and three armoured divisions to move between 16 and 45 miles as the crow flies across the army area from east to west. A major complicating fact was that, of the armoured divisions, two were still east of the Orne. As this lateral passage of lines was planned and conducted in haste, it was a considerable organisational challenge. To assist the conduct of the first leg of the journey, the N13 Caen–Bayeux Road was closed to all other east–west traffic, but not as we will see to essential crossing traffic. The guns of 8 AGRA would lead from the Giberville area east of the Orne.

Breaking off all other now redundant planning, Brigadier Harvey's staff of 29 Armoured Brigade worked in the time available to produce very basic movement instructions, but such was the rush that the usual orders and associated tables could not be typed or disseminated for a move that was to begin at 2300 hours. As usual, movement was conducted under radio silence and the white knight of VIII Corps was removed from the corps troops' vehicles and soldiers' battledress. General Roberts noted that overnight 28–29 July the route crossed:

> … the supply lines of both 12 and 30 Corps. Certain times were given us when we would have priority on the roads, but in fact no one else took any notice of them and we had particular trouble with some of 30 Corps, who professed to be quite unaware of any priority timings for us. Divisional HQ finally came to rest at 2100 hours on the 29th:

After the night's march, the brigade war diary recorded that the: 'Bde began to arrive in new area at 0700 hours. Tks all in by 0900 hrs. Owing to great traffic confusion en route, wheels were not in until 1500 hrs.' The confusion with XXX Corps was a problem that was exacerbated by the route south-east to the

At the time, the N13 was the main artery west for redeployment, running through the village of Vaucelles. Today the village is quieter with the new N13 bypassing the centre

The VIII Corps sign borne on the vehicles and soldiers' shoulders by headquarters and the corps' supporting and administrative units.

Caumont area, which was on narrow, hedged roads through the bocage or rough tracks cut by the pioneers. These were easily blocked by breakdowns and with few alternative routes, the going was inevitably slow. The division's assembly area was between Balleroy and Caumont and for 8 RB in the village of Planquery, where, as Corporal Ault noted: 'The night was spent in a field with vehicles highly camouflaged to avoid detection by enemy aircraft. In actual fact the only air activity was during the night when two or three small bombs were dropped uncomfortably close to the harbour.'

Enemy Forces in the Caumont Area

By this point in the campaign, the four field-grade German infantry divisions had all arrived from the south of France and had relieved all but one of the panzer divisions in front of the Second Army, west of the Orne. The German Army was extremely stretched and according to its own commanders at breaking point, with Hitler steadfastly refusing to countenance any withdrawal. *Generalleutnant* von Drabich-Waechter's 326th Coastal Infantry Division was one of the first formations to be released from the Pas-de-Calais, as the effect of the great deception of Operation FORTITUDE had started to weaken in mid-July. On arrival during 22 July from the coast south of Boulogne, the division had taken over the well-developed defensive positions, complete with extensive minefields, from the 2nd Panzer Division in front of Caumont. Of their three regiments, two were in the line, both with all three of their battalions holding a front of approximately 12 miles.

General von Drabich-Waechter GOC 326th Infantry Division.

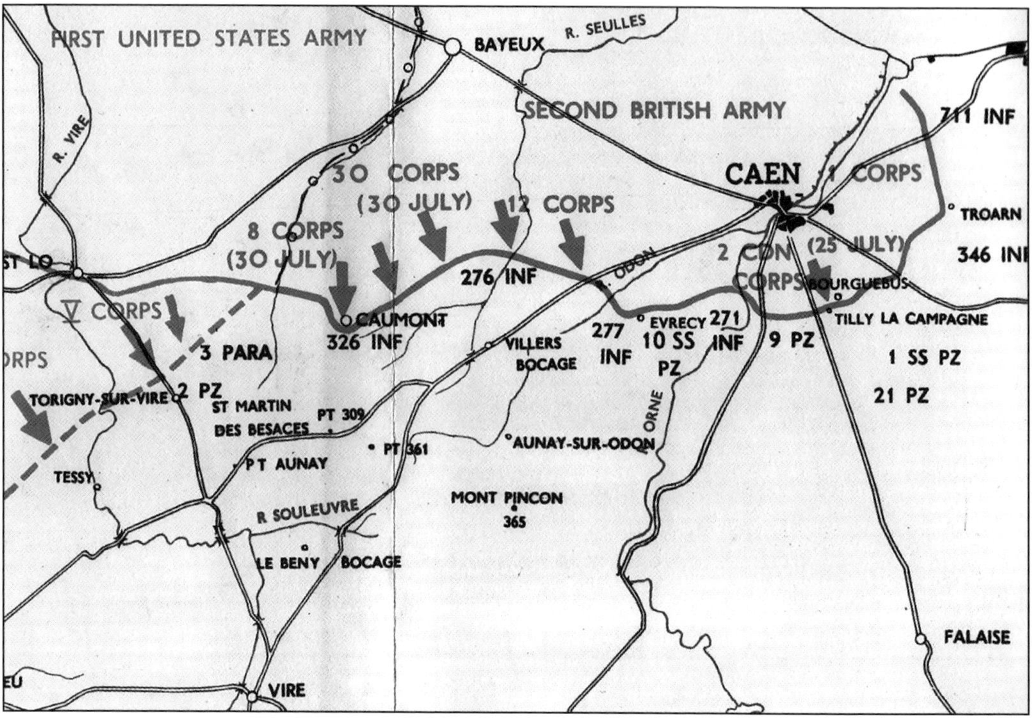

The location of German divisions at the end of July 1944 and the deployment of the British corps for Operation BLUECOAT.

With the Germans under pressure to the west from the American advance down the coastline and the First Canadian Army battling south of Caen, there were no reserves facing the right wing of Second Army. It was assessed that it would take two or three days for either the 9th or 10th SS panzer divisions to redeploy to face the 11th Armoured Division's sector of BLUECOAT. Intelligence, however, reported that 'some of the [thirty] 88 mm equpts of 654 H A/Tk Bn [*Jägdpanthers*] will be on our front'.

The divisional sign of the 326th Infantry Division.

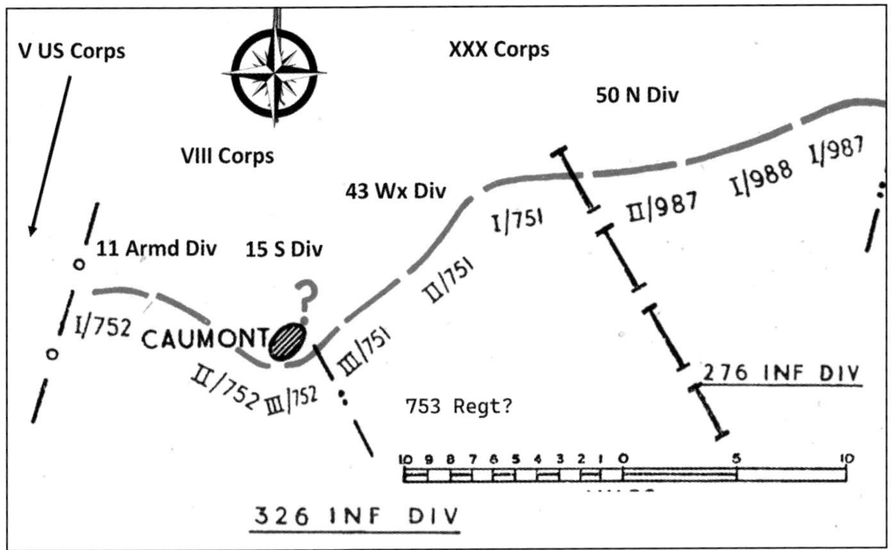

The deployment of the 326th Division in the BLUECOAT area.

Divisional Plans

General Roberts recalled that: 'Corps orders arrived at our HQ during the cross-country march, so that I had, of course, to give verbal orders to the brigadiers at around 0200 hours on 30th July and "H" hour for the attack was 0655 hours. It was a scramble, but just worked.' This is an example of how well the staff, commanders and units were trained and, by now, experienced. The orders process was slick, and the subsequent battle procedure based on standing operational instructions was well understood down the chain of command. The objective of the 15th Scottish Division and 6 Guards Tank Brigade to the left

Ost Truppen and *Beutedeutsch*

Despite Nazi ideology regarding people from the east as *Untermensch*, the Wehrmacht had been taking volunteers from the captured territories since 1941. However, with the losses of 1943 at Stalingrad, 'Tunisgrad' (North Africa) and Kursk, Germany was increasingly desperate for manpower. Consequently, the Germans resorted to wholesale 'recruiting' of Soviet soldiers who were prisoners of war, particularly those from the eastern republics and the reluctant 'Russians' such as the Ukrainians. There were three broad categories of volunteers but in practice they overlapped significantly.

(1) *Hilfswillig*, or 'those willing to help', were known as 'Hiwi'. They were auxiliary volunteers recruited from the people of the eastern territories occupied by Germany. They mainly undertook camp guards, security, administrative or labouring duties to relieve Germans for other tasks.

Osttruppen were usually grouped by ethnicity/nationality. These soldiers are from Georgia and serving in Normandy.

(2) *Osttruppen* or East troops. By 'volunteering' Soviet POWs were able to get out of German POW camps, where there was a policy of deliberate maltreatment of Soviet POWs (in contrast to their treatment of Allied POWs). This policy resulted in some 3.5 million, or over 50 per cent, of all Soviet POWs dying in German captivity. Consequently, in practice it was hard to differentiate between genuine motivation and volunteering for better treatment, plus a greater chance of survival.

By 1944 most infantry divisions, including the 326th, had a pair of *ost* battalions in their order of battle, usually recruited from a particular ethnicity. With a complement of German officers and NCOs, some *ost* units were trusted but all were told that if captured they would be shot by the Allies.

(3) *Beutedeutsch* or 'booty Germans', was the derogatory term for those conscripted from what were regarded as ethic German areas into the Wehrmacht, which had been incorporated into Greater Germany. They were found across the armed forces and served with a varying degrees of willingness or reluctance. There were, of course, those from these territories also subject to conscription, who regarded themselves, for example, as Poles or French, and were most often reluctant 'Germans'. *Beutedeutsch* tended to serve in low-grade or coastal formations such as the 326th Division, where up to 70 per cent were not ethnic Germans. Typically, officers and senior NCOs would be German, of which there was also a hardcore to keep the rest fighting. One writer summed up the situation as every two trenches of Poles having Germans dug in behind them. The Wehrmacht commanders were under few illusions as to the quality of these troops and they were the subject of

particular attention by political officers. The note for a briefing to the soldiers of 712 Regiment on arrival in Normandy was captured and reads:

(a) You are relieving a SS formation in a quiet sector.
(b) You will be opposed by troops of the 5 U.S. Infantry Division who have suffered fairly severe casualties.
(c) There is no armour against you.
(d) The enemy will drop, or fire over, leaflets. No one below the rank of major will pick them up. These leaflets invite you to desert, and enjoy the amenities of British prisoner of war camps. Remember that if you do, you will be taken to England and run the risk of death by V1. After that you will be shipped to the United States or Canada for life-long labour. Or you may be exchanged for a British or American paratrooper (of which we have captured thousands) tried by court-martial and shot. In any case, your families will suffer in consequence, exclusion from the German *Volksgemeinschaft* being the least of their troubles.

By and large these threats did not work, with Germans believing the British and US psy ops material rather than the German.

was the dominating Point 309, which overlooked Saint-Martin-des-Besaces. The 11th Armoured Division's task that General Roberts briefed to his brigade commanders was to protect the Scottish infantry's right flank and at the same time cover the gap west to V US Corps, whose left boundary was the River Drôme. In short, the division was to keep abreast of the advance of the flanking formations.

The plan was for the division to advance on two axis, one per brigade. Any thoughts of an advance across country led by tanks would have been dispelled as they drove into thicker and thicker bocage country as they travelled southwest from Bayeux to their assembly areas. For the advance down a corridor 4,000–5,000 yards wide between the Americans and the Scots, there were few good roads, except the Caumont–Saint-Martin-des-Besaces road, otherwise rarely did they head conveniently south. Consequently, the brigades' axis zigzagged considerably.

On the right 159 Brigade led off with 1 Hereford supported by 2 F&F Yeo on the axis via Cussy and Dampierre, and on the left, 29 Armoured Brigade led with 3 Mons and 23 Hussars, on an axis via Sept-Vents and Saint-Jean-des-Essartiers, 8 RB and 3 RTR. They were to take over the lead once the breakthrough had been made. The logic here was the greater bayonet strength of 3 Mons was better for a fight through and the superior mobility of 8 RB should be employed in exploitation, once through what was assessed as a defensive crust.

The Bocage

The bocage ... is ill-suited to the arts of war. There are no wide expanses for tactical deployment, no scope for manoeuvre or reconnaissance. The country is difficult for infantry and vastly unpromising for tanks, and to

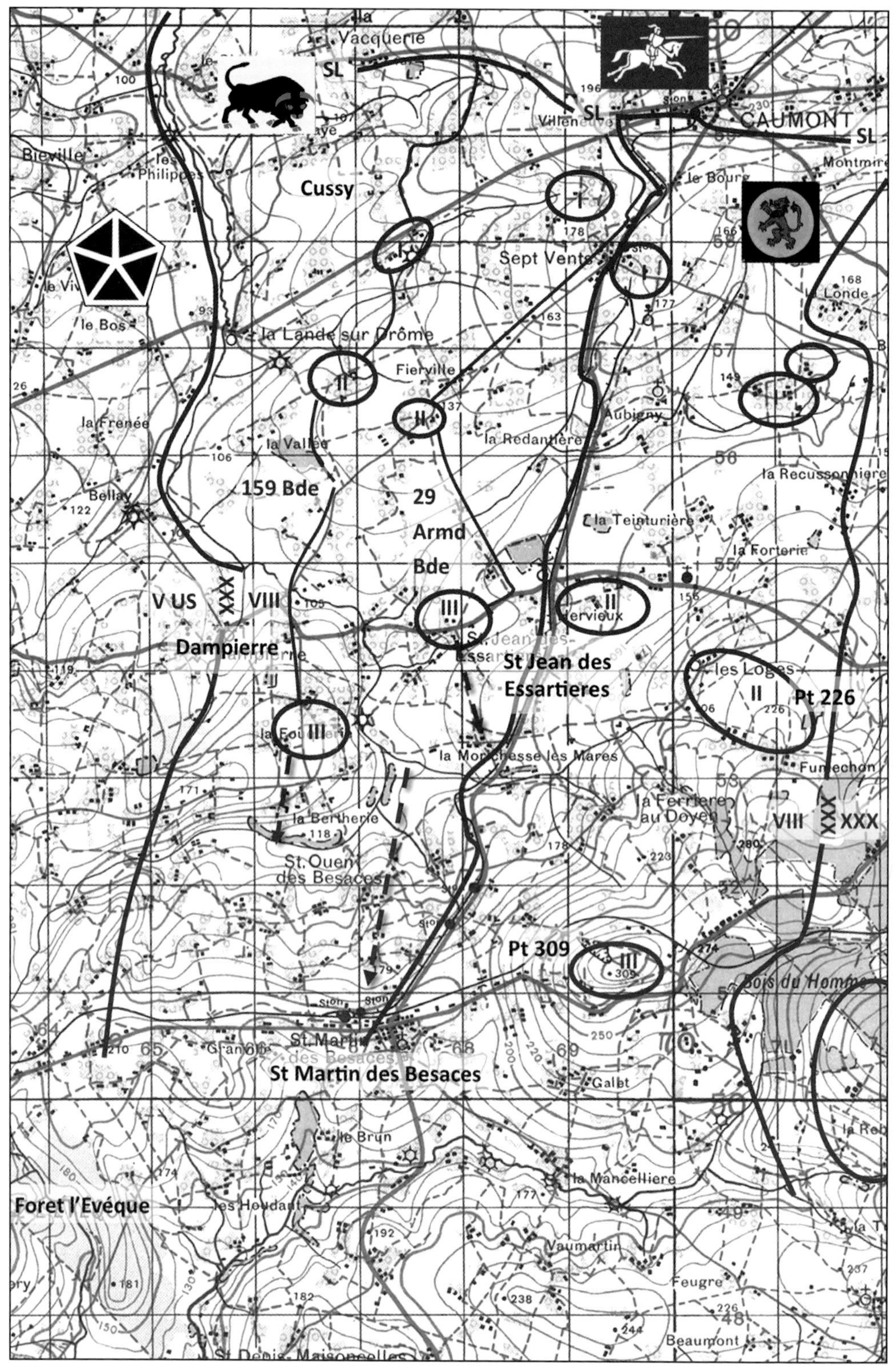

The phases in the divisional plan for the advance to Saint-Martin-des-Besaces and Point 309.

launch an offensive through it might in other circumstances have appeared most imprudent. No divine or satanic author of cosmic struggles would have staged his Armageddon here. [Taurus Pursuant]

On arrival at Planquery, the commanders of 8 RB received a visit and briefing on the difficulties of fighting in the bocage from an officer of 24 Lancers who was well known to the battalion. His briefing was probably along the lines of a passage from one of XXX Corps' 'Immediate Report' issued early in the campaign, which highlighted the advantages for both the attacker and defender in the bocage. For the riflemen, however, 'Any hopes we may have had of an easy passage to come were dispelled.'

This country is probably the best possible country for infantry to operate. They can move out of view of the enemy because of the cover afforded by the hedges and trees in almost any direction. They are a complete menace to the enemy's tanks because they can lie up for them, shoot them at close range, snipe them and make a general nuisance of themselves to any hostile armour. They want to realise thoroughly their superiority, and such expressions as 'being over-run by tanks' have been forbidden, as it is not considered that it is possible for infantry in this country ever to be 'over-run' by tanks. If tanks do come near them they can conceal themselves until a favourable opportunity occurs for them to turn their anti-tank weapons on the tanks or snipe the tank crews. Cleaning up close country of snipers and clearing villages of similar kinds of hostile opposition becomes a frequent and normal task of an infantry battalion; there is nothing difficult or unusual about such operations, but they must be properly run, otherwise people will shoot each other

An aerial photograph of the bocage country south of Caumont.

and not obtain profitable results. Another expression which has been forbidden is 'being mortared out of a position'. Infantry have been heard to complain that they could not hold certain ground because they were 'mortared out of it'. Infantry must realise that when they gain an objective they will for certain receive defensive fire in the form of shelling or mortaring, and they must be prepared for that type of attack. The answer is to take your own shovels and dig your own slit trenches immediately you get on to a position – or make use of the enemy's slit trenches which will almost certainly exist if the objective which you have taken has previously been held by the enemy.

One rifleman summed up the opinion of fighting in the bocage: 'I'm afraid we didn't like it at all; it made us feel rather vulnerable to say the least and it was very difficult to deploy off the road and into the field if we met any serious trouble.' Others, particularly the armoured regiments and those who had fought in the open spaces of the desert, loathed the move west into the hedgerow country and for many of those who had been teetering on the brink, the additional strains resulted in an increase in psychological casualties.[9]

Fire Support

Even though 8 AGRA had been first on the road west, there were few medium guns available for the break-in battle and subsequently there would be a significantly reduced weight of fire from that of EPSOM and GOODWOOD. With fewer barrels, VIII Corps arranged for delivery of 250 rounds per gun of 5.5-inch and 4.5-inch artillery ammunition over and above three days' daily expenditure in compensation. The 25-pounders of 13 RHA and 151 Field Regiment each received 380 rounds. Much of this additional ammunition was held on trucks rather than dumped at battery positions, as it was envisaged that the guns would be moving forward with the advance.

With the rearrangement of the division into battlegroups, the two field regiments remained affiliated to their brigades but with 8 RB now being expected to operate as a battalion, Colonel Hunter received in his tac HQ the BC of 58 battery from 25 Field Regiment RA, attached from 8 AGRA.

Without the accustomed weight of artillery fire, extensive air support was planned, but it was stressed that BLUECOAT would go ahead regardless of whether or not aircraft could fly. The air attacks on depth positions were to be in sequence as the advance progressed, which was one of the lessons of GOODWOOD. It would ensure that the enemy were neutralised as far as possible and with a minimum time for recovery. The plan included sorties by heavy and medium bombers on targets on the division's frontage, close support fighter-bomber attacks by the RAF's 83 Group in front of the brigades and on opportunity targets, armed reconnaissance on approaches to the battle area and tactical and photo recce. Some 500 medium bombers of the Ninth US Air Force were to attack Dampierre between 0855 and 0955 hours, and Point 306 (Quarry Hill) would be targeted by 216 bombers between 1555 and 1655 hours.

Chapter Nine

Operation BLUECOAT

Shortly after 0300 hours, the division's leading battlegroups set out for the start line with the infantry riding on the tanks. For 8 RB, having netted in their radios, rather than blocking the route to the start line, the half-tracks remained parked up in their assembly area in the cover of hedging alongside the tanks of 3 RTR. Major Bell noted that:

> German aircraft were fairly active over the area and the next morning before light came 11 Platoon were bombed in their field. The cause was the lighting of a petrol fire in an adjoining field by another unit, and as was always the case, the punishment descended on the innocent. The bombs, however, all landed right in the centre of the field and the Platoon suffered no casualties.

In the event the battalion spent all morning waiting, as from the outset the normal friction delayed the leading battlegroups' advance, starting with, as already noted, the enemy's extensive use of mines. Due to the late arrival of the 11th Armoured Division's sappers, the intent to provide a 600-yard breach astride both routes could not be met. The mines were so densely laid that the engineers barely had time to clear the brigades' axis by H Hour. Consequently, once again the battle-groups would have to deploy fully once through the minefields. Even so, the 23 Hussars lost several tanks before they reached their start line.

Thanks to cloud cover, the first of the bomber strikes was delivered on a reduced scale, with some aircraft returning home with their bomb load. German prisoners from the 326th Division, most of whom would have endured bombing in the Pas-de-Calais, were adamant that the air attack was 'devastating'. In addition, the headquarters of the 326th Division was hit and among the dead was a senior staff officer. Sorting out the survivors and moving the location of the headquarters no doubt slowed coordination of the German response.

With the infantry following the bombardment and advancing ahead of the tanks, the first phase of BLUECOAT was no armoured charge. The Herefords met strong resistance on their axis at Cussy and despite the defences of Sept-Vents being mainly in 15th Scottish Division's area, 3 Mons and the 23 Hussars faced stiff opposition west of the village. The tanks in particular found moving tactically through the thick country difficult, but 3 Mons and the 15th Division had subdued Sept-Vents by around midday after 'a fairly severe battle'. Though they were through the crust of defences, 23 Hussars were held up by a further series of minefields, but as their historian commented 'very shortly the whole

Carriers of 8 RB parked alongside hedges near Sept-Vents.

group was able to push on south, still moving with the infantry walking in front and still hampered by the bocage'. Meanwhile, back at Planquery, 8 RB had been waiting to move all morning. Captain Straker wrote:

> After a succession of orders and counter-orders the Bn. finally left PLAN-QUERAY at 1400 hrs working in reserve with 3 RTR. Movement was fairly slow, and a certain amount of disorder on the narrow lanes was inevitable. During the early evening we passed through two blazing villages, and were subjected to inquisitive visits by enemy planes, fortunately without result.

Despite moving forward, as recorded in the battalion's war diary, they spent the afternoon 'following the successful advance along the centre line SOUTH through CAUMONT and LES-SEPT-VENTS'. Up ahead of them, however, the 23 Hussars' historian reported making:

> slow and steady progress, fighting no serious engagement, but having an occasional brush with enemy infantry and light vehicles. It was growing dusk

when we reached the outskirts of St. Jean d'Essartiers and found that it was German-held. B Squadron engaged the enemy frontally, while C Squadron came in from the eastern flank and their Third Troop knocked out a 75-millimetre gun as they did so. The opposition then collapsed and we entered the village in almost pitch darkness except for one enormous fire which kept shooting sparks all over any tank that passed it.[1]

As it was getting dark, that evening 8 RB took over the advance for the last 3 miles to Saint-Martin-des-Besaces, with G Company leading and 3 RTR following, but after less than a mile they encountered the enemy. Major Bell wrote:

Shortly afterwards the Company took the lead and, pushing south, arrived before a small village [la Morlecheess le Mares] just as dusk was falling. Enemy had been observed in the village and we were ordered to clear it as quickly as possible. The gunners, having laid on plans for an impressive 'Stonk' on the place, informed me just before the attack was due to begin that they could not get permission to fire as higher formation did not think there were any enemy there; and, having no time to argue as darkness was fast coming on, I made a hasty arrangement with the troop leader of the 23rd Hussars [*sic* – 3 RTR] to pump some shells into the village and the attack went in as previously planned, being carried out by 10 and 11 Platoons. In the village we found enemy equipment which had obviously been very recently abandoned, and one wounded German. However, it was very dark by now and it was useless trying to search the houses, so we spent the night very much on the alert and wondering what the morning would bring forth.

As darkness fell, the battlegroups of 29 Armoured Brigade, having advanced 8 miles, were beginning to harbour for the night, including 23 Hussars:

Our entry into the village had been from two directions, and what with this, the darkness, the fire and the narrowness of the lanes, a traffic jam of considerable dimensions had to be sorted out before the Regiment and the Battalion [3 Mons] eventually crammed themselves into a very small field for the night.

As he recorded, General Roberts was of a similar mind about stopping for the night:

After a very tiring approach march and little sleep and then a hard day's fighting, there was a natural tendency 'to call it a day as soon as the sun went down', but firm orders came from the Corps Commander that we could not relax.

The Churchills of 6 Guards Tank Brigade, after a remarkable advance, had seized Point 309 and were later joined by the Scottish infantry. It overlooked Saint-Martin-des-Besaces but, being a mile distant, they could not dominate the

Major General Roberts mounted in his White scout car in Normandy. This vehicle was more convenient than his tank for visits to headquarters and better protected than the Jeep he typically used in the rear area.

Germans in the village below. But, as General Roberts explained, it was a key objective that had to be taken quickly:

> The main village in front of us was St Martin-des-Besaces; it was an important cross-roads; one main road ran east–west and the other north–south. We could hold it in front and attack from the west; this would be the least-costly arrangement. Accordingly, instructions were issued to this effect.

Saint-Martin-des-Besaces

The village of Saint-Martin-des-Besaces was held by a recently arrived *kampfgruppe* based on I/125 *Panzergrenadier* Regiment of the thinly spread 21st Panzer Division that had been deployed rapidly that day to bolster 326th Division's

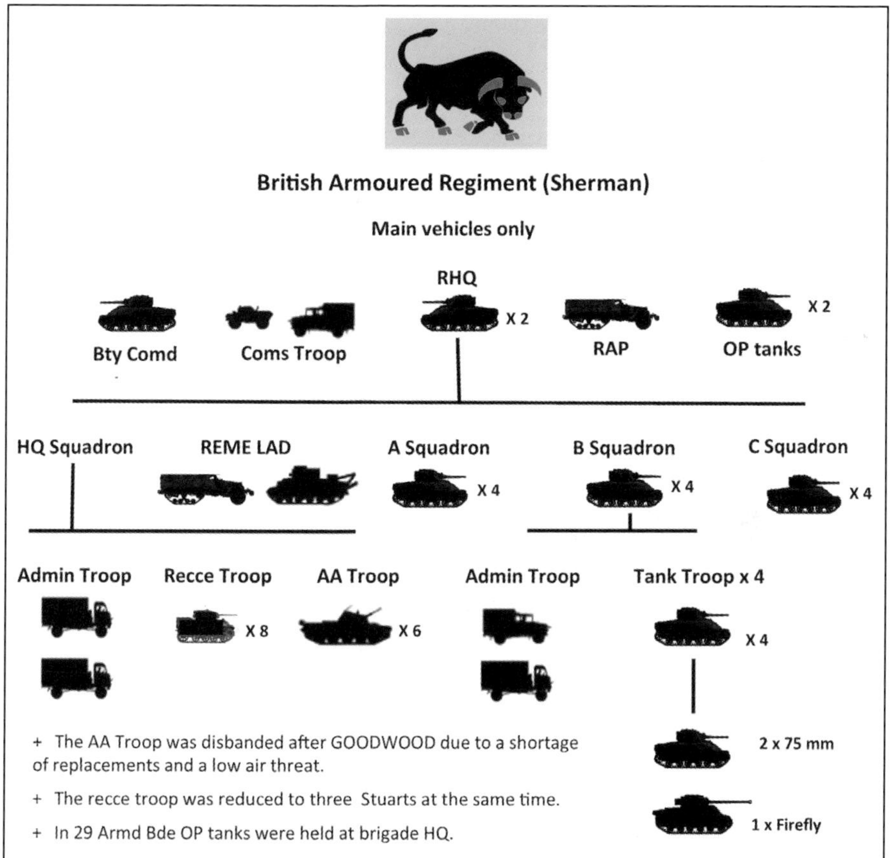

The British Sherman-equipped armoured regiment.

crumbling front. It probably numbered at the most 300 infantry along with support weapons.

The war diary entry for that evening made by a staff officer in headquarters 29 Brigade reads: 'During the night 30/31 Jul 8 RB were ordered to attack ST MARTIN DES BESCES and clear it if possible by first light.' This would require an advance of some 3 miles to the hilltop village. Also tasked to assist was 159 Brigade. General Roberts continued with his account of that night's operations:

[Brigadier] Jack Churcher was told to arrange for the 4 KSLI, who were the most forward regiment, to establish themselves during the night astride the road leading west out of St Martin. Churcher said this was impossible and, perhaps rather weakly, I said that the Corps Commander had ordered it and it had to be done. I think he said much the same to the CO of 4 KSLI, Max Robinson, and so the attempt was made.[2]

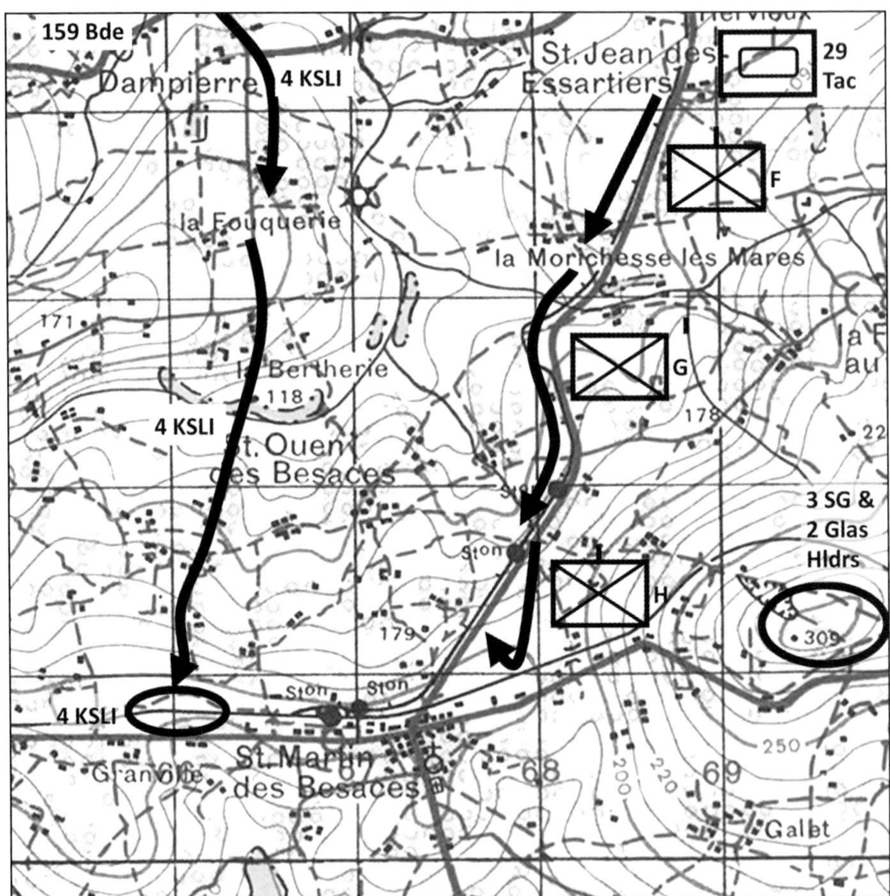

The fighting advance of 8 RB to Saint-Martin-des-Besaces and the infiltration of 4 KSLI.

The 8 RB's advance was initially led by F Company on foot following the road and railway line from Saint-Jean-des-Essartiers, their history recording that 'an exciting journey in the dark ensued, with occasional patrol clashes'. Having secured the two railway stations at the foot of the ridge, Captain Straker and H Company took over for the last 1,000 yards uphill to Saint-Martin:

> The first part proceeded according to plan and 14 Pl were duly sent forward to see if the village was clear. A slight delay caused by the arrival of the C/Sgt arriving apparently from the direction of the enemy with, to his obvious pride and joy, fresh bread and meat (the first of the campaign) met with none of the praise it deserved owing to the task on hand. 14 Pl went forward on foot towards the top of the hill overlooking the village and met opposition in the form of a tank tucked away on the road under a steep bank. Deciding the village was held, 14 Pl eventually withdrew but not before Cpl. Fulton had

considerably shaken the occupants of the tank with a well-aimed grenade, causing it to disappear back in the direction of the village. 15 P1 moved forward in support and the two Pls. remained for the rest of the night on the reverse slope of the hill.

H Company was held up unable to cross the tall east–west railway embankment just to the north of the outskirts of Saint-Martin. Meanwhile, 4 KSLI, who had been stirred from sleep, were on the move at 0300 hours. They were fortunate enough to find little-used tracks and roads along which they made their way in single file, without encountering any Germans. They established themselves astride the railway 1,000 yards west of Saint-Martin.

At 0015 hours Colonel De Winton, the Deputy Commander of 29 Brigade, held orders at Saint-Jean-des-Essartiers for the following morning's attack on Saint-Martin. Colonel Silvertop, CO of 3 RTR, allocated C Squadron to the close support of 8 RB's attack from the north-east and B Squadron to take up positions directly north of Saint-Martin to provide fire support. A Squadron was in reserve.

As dawn broke, H Company resumed the attack on Saint-Martin but with the German infantry I/125 *Panzergrenadiers* positioned in the paddocks, orchards and buildings, in attempting to cross the railway line the riflemen made easy targets for the enemy defences in and around the village:

At first light an attack was put in by the Company, but was held up by heavy Spandau fire from well-concealed positions in the cornfields. Several men were killed or wounded including Capt. Straker, who, mistaking a German sniper for a member of the Company, ordered him to put on his steel helmet. The answer was a burst of fire which fortunately caused only slight wounds. Capt. May then came forward to take over command. Meanwhile, G Coy. had moved up on the left flank, and put in an attack, but were also held up . . .

The attack on Saint-Martin on 31 July 1944.

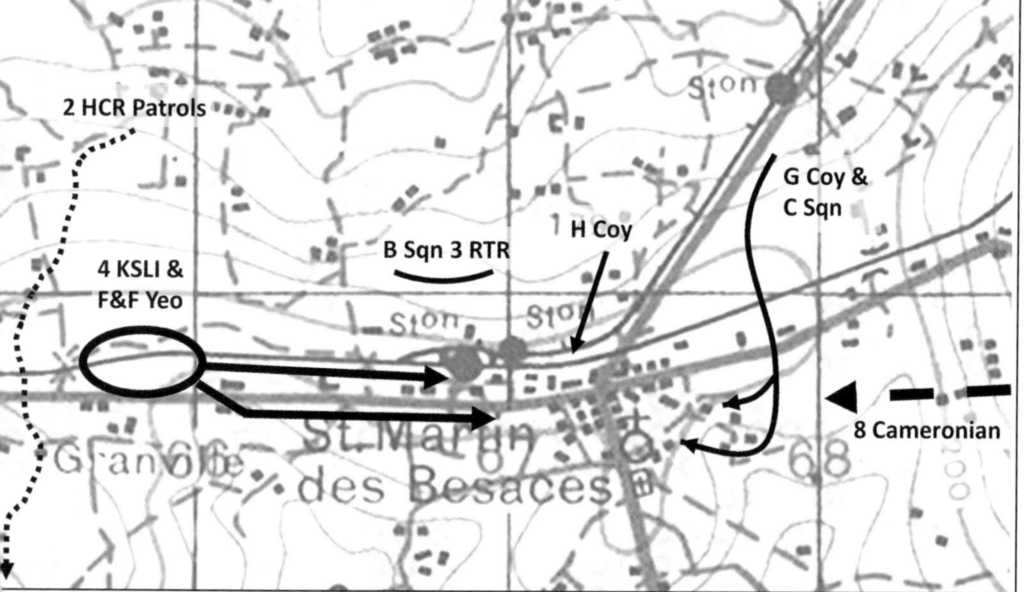

Major Bell wrote:

> 'H' Company had passed through us during the night, and in the morning, having searched our village and found nothing, we pushed on toward St. Martin des Besaces, where 'H' Company were in trouble and had been pinned down. We were ordered to attack, supported by a squadron [C Squadron] of the 3rd R.T.R., being given the information that the enemy consisted only of the odd sniper, and the three motor platoons accordingly went into the attack, the carriers being held in reserve. A high railway embankment had to be crossed and this obstacle could not be surmounted by the tanks and so they were unable to support the motor platoons. 10 and 11 Platoons came under heavy fire and were pinned down, suffering heavy casualties, but 12 Platoon managed to work round and reached their objective, although they lost a complete section on the way through mortaring. For several hours this handful of men of 12 Platoon held out on their objective against heavy odds, and Eric Yetman, who, although wounded, refused to go back, was awarded the Military Cross for this action … St. Martin des Besaces had proved to be a much tougher nut to crack than higher command imagined, and the whole operation lacked co-ordination.

Also out on this flank 15th Scottish Division's recce regiment had been unable to work their way across the railway line just east of Saint-Martin.

The bold attempt by 8 RB to attack head on a strongly held position encountered during exploitation, across difficult ground, cost the riflemen nine killed and twenty-one wounded. As a result, G Company joined F Company in reorganising with two motor platoons. '11 Platoon had to be disbanded and split amongst 10 and 12 Platoons.'

The war diary entry by 159 Brigade on the morning of 31 August clearly reflects the difficulties 8 RB was having in the village. With 8 RB fixing enemy attention to the north, 4 KSLI was to attack from the west but: 'This action was cancelled twice and plan changed twice before being finally executed owing to our own shelling.' Saint-Martin was successfully attacked, with the Germans apparently unaware of the presence of 4 KSLI, who with the close support of the F&F Yeo, on far more favourable ground, made steady progress. The village was occupied by 1100 hours, with G and H companies being able to advance once the KSLI were in the western part of the village, but Saint-Martin was only deemed clear by 1500 hours. However, General Roberts later wrote that: 'This important success was completely eclipsed however by an astonishing stroke of good fortune …'

The Advance Continued

While the protracted fighting was under way in Saint-Martin, the armoured cars of C and D squadrons of 2 Household Cavalry Regiment were attempting to slip past the village. C Squadron was successful to the west and were fanning out down the south side of the ridge. Lieutenant Powle's patrol of two Daimler

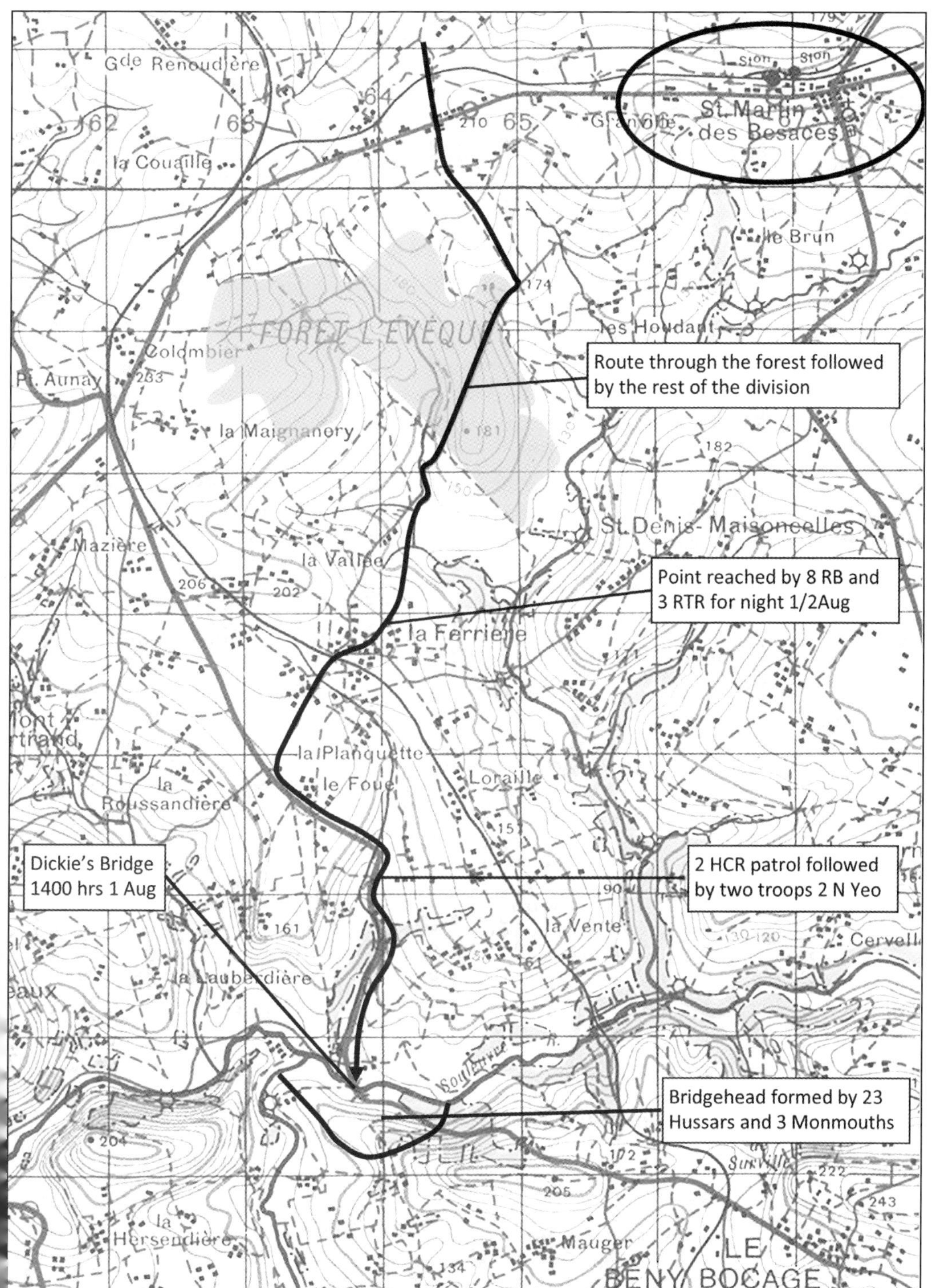

The patrol to 'Dickie's' Bridge via the Forêt l'Évêque by Lieutenant Richard Powle.

armoured cars and two scout cars drove over a mile to the Forêt l'Évêque and there they found the road through the forest almost unbelievably unguarded by the Germans. It was the boundary between 326th and 3rd *Fallschirmjäger* divisions, which was confirmed by a radio intercept of an angry conversation, full of mutual recrimination, between the two divisional commanders. In fact, it was not just a road that was undefended but a thinly held gap of almost 7 miles between the main defences of the two divisions! 'Later it was put to General "Pip" Roberts by one of his staff officers that he could have been held up in the hilly, thickly wooded country by "two men and a boy". "Yes, Joe," he replied, "but they didn't have the boy."'[3] Such are the fortunes of war that were certainly not being enjoyed by XXX Corps, who were facing sterner resistance to the east.

This 'good fortune', as described by General Roberts, 'opened up the front to an even greater extent and further increased the tempo of the advance', which was significantly aided by Lieutenant Richard Powle continuing to probe south, not just the 2 miles to the southern end of the forest but 5 miles to secure an undefended bridge over the River Souleuvre at about 1400 hours. On a 'difficult' radio link, this success was duly reported and General Roberts commented that:

> This was wonderful news and I obviously had to do something about it quickly … But we wanted something down at the bridge at once; the Northamptonshire Yeomanry, working with the KSLI, were the nearest tanks available; six tanks (two troops) of the Northamptonshire Yeomanry were flank protection west of Saint-Martin and these were ordered to the bridge along the same route as that taken by the troop of the Household Cavalry.

With the bridge more secure, Roberts recalled that he:

> … needed a real 'thruster' to get to this bridge 5 miles away. The 23rd Hussars/3rd Monmouths group was comparatively fresh and Perry Harding, commanding 23rd Hussars, was certainly a 'thruster'. I contacted Roscoe Harvey and told him to get this group organised as soon as possible.

News of the leap south that the 11th Armoured Division was making had reached Headquarters VIII Corps and shortly after 1400 hours General O'Connor issued a 'special directive'. 'For the future development of the Allied plan it is important that:

(a) You capture the high ground round Point Aunay to-night [159 Brigade], so that the advance to Etouvy can proceed rapidly to-morrow.

(b) You should occupy to-night Point 204, Point 205 and also the high ground east of Le Beny Bocage about Point 266. The latter point will be occupied to-night in sufficient strength for the bridges over the River Souleuvre to be seized from the south to-morrow morning. 2 Household

> Cavalry will to-night patrol south with the utmost vigour in the direction of Vire. This task is also essential to the Allied plan.

Meanwhile, 8 RB had been relieved in Saint-Martin by 8 Cameronians of the 15th Scottish Division and by 1600 hours was driving south with a squadron of 3 RTR heading for the Forêt l'Évêque, following 23 Hussars. As noted by Sergeant Hicks, despite the earlier absence of Germans, there were plenty still to be rounded up:

> We pushed on later that afternoon through the Forêt l'Évêque, where we gathered up a large number of German prisoners and for about half an hour, I had the pleasure of guarding a high-ranking German officer on my carrier until I could hand him over to be escorted back down the line. He had been relieved of his Luger pistol and he just sat in the back of the carrier, unsmiling and poker-faced and I thought a little bewildered at finding himself in the undignified position of being a prisoner-of-war.

F Company commented that:

> When this village [Saint-Martin] was finally taken, we realised that we had achieved our first real break-through, and the fruits of two months' hard fighting suddenly blossomed forth. The Battalion was now fighting in support of the 3rd Royal Tanks, and we raced on together through the depths of the beautiful Forêt l'Évêque. At last we knew what it meant to be swanning.

With the events to be recorded in the next chapter, this statement was perhaps a little premature!

During the move south, 8 RB met a column of US tanks coming in the opposite direction. As recorded by H Company, this 'caused many of the Coy's drivers to think that their last hour had come'. Major Bell remembered G Company's arrival in La Ferrière, 'where the civilians gave us an enthusiastic welcome, and no enemy opposition was encountered on the way'. This is where 8 RB and a squadron of 3 RTR halted for the night, with 23 Hussars and 3 Mons holding a bridgehead across the River Souleuvre. General O'Connor's orders, typically seeking to exploit the situation, had come too late in the day to fulfil, but a day that had looked likely to be one of fighting around Point 309 and Saint-Martin ended with the 11th Armoured Division having advanced over 5 miles into a salient.

Once halted for the night, the battalion was able to enjoy being 'liberators' and, as noted in F Company's account:

> At La Ferrière we linked up with the Americans, and by now we were receiving a rapturous welcome from the French civilians. Calvados and cider were offered in abundance, were drunk with caution by those who knew, and not so much caution by those who didn't, and we realised that war possessed a pleasant side.

The division's infantry riding on the Shermans of 29 Armoured Brigade. After BLUECOAT it was decided to use TCVs wherever practical as the process of mounting was so slow.

As described in the battalion's war diary, the encounter with the column of V US Corps' tanks was only the first such meeting:

> We were now through enemy's gun line into country from which he had not had time to evacuate the civilians, as a result we spent a most amusing and sociable 12 hours in PERRIERES [*sic* – La Ferrière] with not only the French population but also a Bn. of American Infantry in whose area we now were, who came through us to take up a posn in the area. The close liaison between the two Armies was symbolised by the action of an American enlisted man who, as he marched through the village, silently and without stopping, thrust into the hand of the Colonel a stick of chewing gum.

G Company's stay in La Ferrière had been short before being ordered forward another 1½ miles to form an outpost:

> After a brief period of rejoicing we moved forward again with a squadron of the 3rd R.T.R. and reached St. Charles de Percy, thus cutting the main highway from Caen to Vire. The houses on the crossroads were cleared without much difficulty, and prisoners were taken, in addition to a collection of enemy soft vehicles which were successfully brewed by the tanks.

The German Reaction and a Change of Plan

The coastal infantrymen of 326th Division had not stood up well to the bombardments by artillery and air that was followed by the onslaught of VIII Corps and had all but collapsed. However, against field-grade infantry divisions XXX Corps was making only slow progress, not helped by the arrival of the leading elements of 21st Panzer Division, which were also coming into action on the exposed left of 15th Scottish Division. However, on the evening of 31 July tactical air recce reported 'large scale armoured movement west'. This and the likelihood of a counter-attack on 15th Scottish Division's flank, led to the release of the Guards Armoured Division during the afternoon, which was to come up on the left flank of the 11th Armoured Division for the advance south. From the scale of enemy armoured movement, British commanders rightly concluded that the 21st Panzer would be followed by the divisions of II SS Panzer Corps.

Overnight 31 July–1 August a departure from the original plan formulated by Second Army was agreed. VIII Corps' history recorded that:

> ... in view of the failure of 30 Corps to achieve an equally satisfactory advance, and on hearing of this latest success by 8 Corps, the Army Commander had decided that the latter was to become the main striking exponent in the operation. He accordingly assigned to it the objective originally allotted to 30 Corps in this area, namely the commanding heights around Le Beny Bocage [as identified by General O'Connor]. The Allied swing to the east was beginning even earlier and further north than had been anticipated, for 8 Corps was now to converge on to the axis of advance of 30 Corps.

Generals O'Connor, Dempsey and Montgomery. With relatively short distances between headquarters in the tight lodgment, such meetings were a regular feature.

Le Bény-Bocage

Having spent the night with a squadron of the 3 RTR covering the road south of La Ferrière, G Company was 'standing-too'[4] at dawn on 1 August, 'when our area was sprayed by machine-gun fire coming, we later discovered, from an enemy tank which seemed to be rather out of the picture and which withdrew hurriedly when it realized what it was up against'. The advance was resumed at 0650 hours, with F Company leading with a squadron of tanks. They covered the 2 miles to the River Souleuvre and the bridgehead, then advancing another 2 miles before swinging north to Point 243, which overlooked Le Bény-Bocage. Their account demonstrates that the enemy were still around:

> On we went to Le Beny Bocage, where we had a sharp encounter before seizing a piece of high ground to the north of the town. From here No. 1 Section of 5 Platoon was sent out on patrol to the north, but was shot up by a camouflaged vehicle, either a tank or self-propelled gun, and casualties were suffered. This reminded us that there was still enemy about, and that the battle was very much on.

G Company had missed out on the previous evening's liberation celebrations but were now the first into Le Bény-Bocage. Rifleman Kingsmill described the French people's rejoicing:

> We had a really wonderful experience this morning. If I may use the word, we helped to 'liberate' a fairish-sized town. I have truly never before seen people go so absolutely mad with joy. Everybody was either shouting, waving, cheering, clapping, kissing one another, singing the 'Marseillaise' or doing all the whole lot at once. I wonder if I can give you the picture.

The advance on Le Bény-Bocage during 1 August 1944.

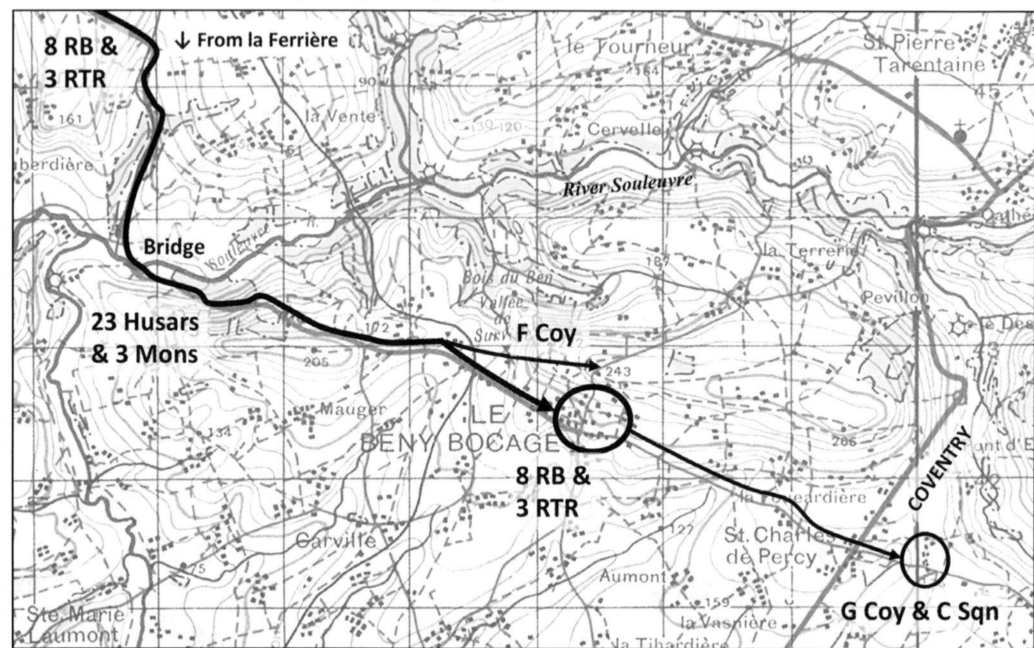

The town square market-place, I suppose, a fairish-sized rectangle about 80 yards by 20 yards. At one end a Jerry tank, charred and blackened, still pours forth smoke. At the other end, where I am, there is a party going on in a hotel, you can see them from the open (very open – no glass) windows. Any soldier who can be 'spared' for a moment gets raked in.

I, with headphones round neck, must perforce stay in my truck, but the back is 'open' and I can stand up and view the scene. A man comes up, smiling from ear to ear, and gives me a glass of champagne. A lovely drink it was, cool and bubbly; but before I know it my hand is seized by an old lady – about eighty-five, I should think – who shakes it madly and cries, 'Merci monsieur, merci beaucoup.' She is nearly weeping for joy. It kind of gives you a lump in the throat to witness it. A gendarme is running up and down the square waving a huge tricolour, and more tricolours are flying everywhere – from telegraph poles (how the hell did they get there?), one stuck in the fountain; most everybody has one some-where in or around his house. Fresh 'Cross of Lorraine' armbands appear on young men, who by their look are proud to be Frenchmen again, and not just lackeys. Another glass of wine is offered (and accepted), and more follows that. Flowers are thrown at you – nearly every truck has flowers on it. Then it's our turn; we give the people sweets. Some bars of chocolate and some cigarettes – no charity, it's a real pleasure to give them.

At this stage in the campaign there was, however, occasionally a darker side to liberating villages:

During the advance there were unfortunately all too many occasions when we passed through scenes of passionate excitement and wildly waving flags, and no sooner had we gone than the Germans appeared while the demonstrations of joy at their departure were still in progress.

G Company's stay in Le Bény-Bocage had been short before being ordered forward another 1½ miles to form an outpost.

After a brief period of rejoicing we moved forward again with a squadron of the 3rd R.T.R. and reached St. Charles de Percy,[5] thus cutting the main highway from Caen to Vire (COVENTRY). The houses on the crossroads were cleared without much difficulty, and prisoners were taken …

Out on a limb at the tip of a salient deep in enemy territory, G Company group realised they were vulnerable:

It was evident that we had cut one of the enemy's main escape routes from the north, and we anxiously waited to see what would be his reaction to our move. We settled down astride the crossroads and it was not long before an enemy tank approached from the north and we could see movement of enemy armour to the south. Our chances of holding off an armoured attack in any strength were not bright with the forces we had at our disposal, so after urgent appeals for help over the air, some self-propelled 17-pdrs were put under our command and we were further strengthened by the loan of a motor platoon from 'H' Company. The expected counter-attack never came but the threat of it kept us all very keyed up for the rest of the day and night, our only aggressive action, however, being taken by the mortars in the form of a very good shoot on enemy movement on the Vire road to our south.

We laid Hawkins grenades freely across the roads leading to our positions, and the next day began with an unfortunate incident when an armoured car of the Inns of Court Regiment[6] was blown up on a grenade we had failed to pick up. Not a great deal of damage, however, was done to the car, and the occupants, though slightly shaken, were not seriously injured.

At a junction on COVENTRY near Le Bény-Bocage, 11th Armoured Division's Shermans and infantry with German POWs seated on the verge.

Overall Situation

To the east of 11th Armoured Division, the Guards Armoured Division made slow progress during 1 August thanks to stiff opposition and were still 2 miles north of the River Souleuvre at nightfall. In cavalry circles comments were beginning to be made that they were showing too much of their infantry heritage in their slow, overcautious movement but arguably they were facing sterner resistance. Nonetheless, General Roberts was ordered to continue the advance the following day.

Now VIII Corps was holding the main effort, the 11th and Guards armoured divisions were to press on to their objectives. To achieve this General Roberts ordered a regrouping of 8 RB with 23 Hussars. For the advance to the Souleuvre during the 31st, 3 Mons had been mounted on the Shermans and any other vehicle in the battlegroup capable of carrying them. As a short-term expedient,

One of 3 RTR reserve squadron's tanks passing through Le Bény-Bocage. With the commander keeping his head down, there was clearly still a sniper threat.

it had worked well but for a speedier advance and to maintain tactical balance across the battlegroup, as they thrust south against expected enemy resistance, the motor battalion mounted in its own half-tracks was a far better option for spearheading the division. Of this relationship, the Rifle Brigade's regimental historian observed:

> The Battalion worked with the 23rd Hussars, a connection which was to last, with various short periods of separation, until the end of the war. No armoured regimental group could have worked together with more harmony or more effectively. Squadrons and companies, troops and platoons, came to know each other perfectly. When: the armour was the dominant arm Colonel Perry Harding commanded: as soon as the tanks were held up, Colonel Tony Hunter automatically took over. It was because of the mutual confidence established between the two that joint operations could be laid on quickly with hardly any need for the pause: and waits and order groups and preparations, which can readily devour the hours of daylight in the face of the enemy.

This relationship extended down to the commanders of the motor companies and squadrons, who typically worked together along the lines of colonels Hunter and Harding. The problems of GOODWOOD's second day had been resolved in both theory and practice.

11th Armoured Division's traffic near the Souleuvre bridge, 2 August 1944.

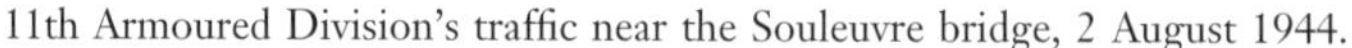

Chapter Ten

Le Bas Perrier

The division was to resume its advance on 2 August on two axis, with 159 Brigade being led by the F&F Yeo and 3 Monmouths, while 2,000 yards to the east, working in company squadron groups, 29 Armoured Brigade's column was headed by 8 RB and 23 Hussars. The mission was to advance the 10 miles south to cut the next main artery, the Vier–Vassey road (RUGBY). The route 8 RB was to take was via Le Désert, Presles, the Perrier Ridge, Chênedollé and on to the high ground overlooking the main road.

In his capacity as the H Company's temporary Anti-Tank Platoon commander, Sergeant Hicks attended Captain May's orders with B Squadron:

> Very early on the morning of August 2nd the usual 'O' Group was called to discuss the planned objective for that day ... The formation of the tank squadrons who were to lead the advance and to protect the flanks and ourselves in the 'softer' vehicles forming the core was decided upon. The centre line and the intermediate objectives were all sorted out and noted on our maps.[1]

The advance was resumed at dawn, 0600 hours, with B Squadron and H Company passing through G Company's overnight positions on the Vier–Villers-Bocage road (COVENTRY). Within half an hour, the Shermans had spotted a German recce patrol and as they withdrew, Sergeant Williams knocked out a pair of armoured cars, which were identified as belonging to the 9th (*Hohenstaufen*) SS Panzer Division. These were the leading elements of the division advancing west towards Vier and would clash with the 11th Armoured Division as the day and the advance progressed.

The German plan was to relieve II SS Panzer Corps from the area immediately west of Caen with an infantry division and send its two formations, the *Hohenstaufen* and 10th *Frundsberg* SS panzer divisions, marching west to confront the BLUECOAT/COBRA breakout. *Obergruppenführer* Bittrich's intent for his corps was to secure the important road junction town of Vier and make contact with II *Fallschirmjäger* Corps, before swinging back east to envelop the 11th Armoured Division's penetration and destroy it. The *Hohenstaufen* was first on the move, broadly divided into two *kampfgruppen*, the northerly of which would counterattack the Guards Armoured Division's thrust, while further south *Kampfgruppe* Weiss headed for Vier. This *kampfgruppe* was based on *Hauptsturmführer* Gräbner's 9th SS *Panzeraufklürunge Abteilung* reinforced with Panthers of I/9 SS *Panzerregiment* and the remaining Tigers of Weiss' 102 *Schwere* Panzer

Sturmbannführer Hans Weiss, commander of the corps' heavy tank battalion (Tigers), led the *kampfgruppe* of his name.

Abteilung. It was this force that ran into 11th Armoured Division's left flank and fought at Le Bas Perrier.

The 23 H/8 RB battlegroup pressed on south, up onto the first ridge of high ground (Point 218) and approached the Vier–Aunay road (WARWICK) at 1030 hours, having advanced some 3 miles (see map page 204). The brigade's right flank was secure, with 159 Brigade's battlegroups visible to the right, but on the left the problem was that with the Guards Armoured Division still only making slow progress there was an increasingly long, open flank. To help, 3 RTR, now battlegrouped with 4 KSLI, were to cover the left rear of 8 RB and 23 Hussars.

As Rifleman Jefferson, a member of 17 Anti-Tank Platoon, recalled, on reaching Point 218 a party of German infantry, believed to be nine soldiers of 326 Division and *Panzeraufklürunge Abteilung* 9, were seen probing west:

The narrow road towards Le Bas Perrier ridge caused several delays. Almost half-way up we came to a stop and we dismounted and took up covering positions. To our left we could see German infantry crossing the field

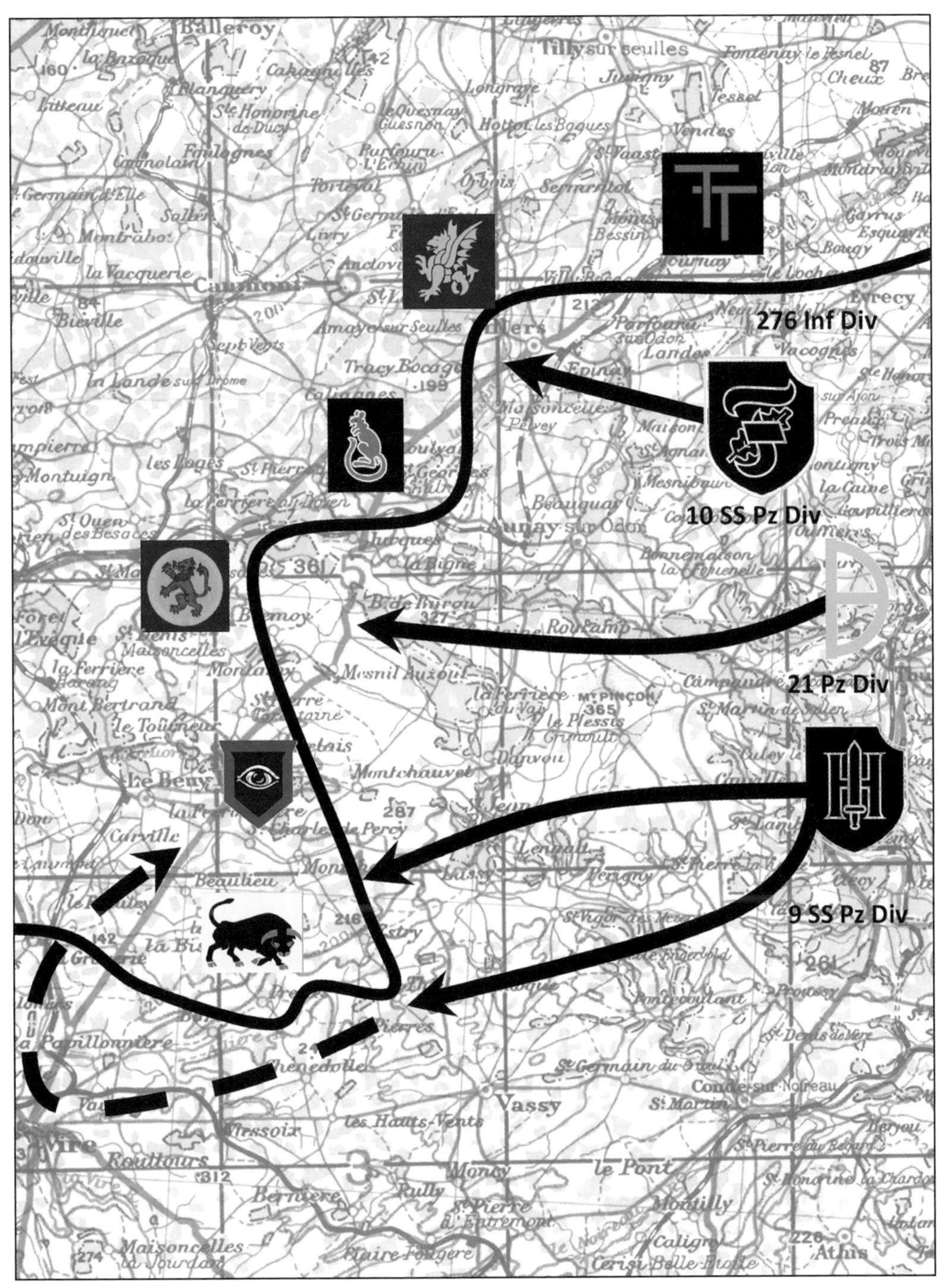

The German counter-stroke against BLUECOAT during 2 August 1944.

> towards us. Hedgerows gave us some cover ... they came forward across the open meadow as if they were on an exercise. At an uncomfortable 20 or 30 yards away we opened fire and many fell dead or wounded.[2]

While covering the left flank, F Company also had an encounter with the enemy near Point 218:

> ... by now we were far in advance of any other British units, and the inevitable enemy reaction followed, for we had penetrated very deeply into his defences. 8 Platoon and No. 2 Section of the Scout Platoon fought a very lively encounter against some machine gun posts on our left flank which had been attacking our Centre Line; these were eventually destroyed, and with the welcome help of a very willing and cheerful AVRE crew we took some prisoners.

Now facing sterner and more numerous resistance, with H Company leading, Rifleman Jefferson describes the nature of the advance:

> Denied scope for manoeuvre, our [supporting] tanks were reduced to the role of blind, slow and highly vulnerable infantry support guns. Their primary task was to knock out enemy machine-gun positions with 75 mm high-explosive shells and these they could seldom see. Consequently, and it was very difficult, we had to devise a means of identifying these cunningly concealed targets for the tank commander. We soon learned not to climb onto the tank and shout. Due to engine noise he could seldom hear and, for us, it proved a lethal pastime. We tried firing Very lights towards the target and also Bren bursts of all-tracer rounds, but neither was satisfactory. The most successful arrangement was for the tank commander to throw out a head and breast microphone set, but even then engine noise made it difficult.

Presles was reached by 1200 hours without further serious incident, amidst great rejoicing from the inhabitants who showed their gratitude, unfortunately short lived, by pressing eggs and potent liqueurs on all and sundry. Later, following behind the company squadron groups, Major Blacker, the 23 Hussars' regimental second in command, was dismounted and directing the sundry elements of the battlegroup at the rear of the column. Sergeant McCully of A Squadron wrote:

> At this point, Major Blacker ... walking across the crossroads and glancing in the direction of Estry [east] was surprised to see a Panther tank. Fortunately, the Panther did not react quickly and Major Blacker was able to take cover. The situation became confused as the high banks at the side of the road precluded either side being able to get a shot at the other. My troop was instructed to work round to a position in the Panther's flank and we were successful in destroying it.
>
> With Panthers liable to arrive unexpectedly, A Squadron was ordered to move out to the left. We followed a sunken lane for some distance until the leading troop was prevented from going farther. At this point, Major

Carriers and half-tracks of 8 RB on the road down from Point 218 to the village of Presles. The Perrier Ridge can be seen rising in the distance and the same spot today.

'Jimmy' Watt, the Squadron Leader, dismounted from his tank and instructed me to take my tank through a gateway into a field and take up a position at the far hedge, where I could observe in an easterly direction for the enemy. As the last of A Squadron tanks entered the field, 'All Hell' broke loose. A number of Panther tanks had been lying in ambush only a few hundred yards away and were camouflaged so well that it was impossible to detect them until they opened up. In no time a number of our tanks were erupting in flames and those that were not were fighting back desperately.

The Panthers mentioned above almost certainly included the Tigers of *Obersturmführer* Kall's No. 1 Company of 102 *Schwere* panzer *Abteilung*, a part of *Kampfgruppe* Weiss. Sergeant McCully continued:

> Orders were given to withdraw and join the rest of the Regiment at Le Bas Perrier. In reversing through the rear hedge, my tank crashed down a sunken lane and, running over a large tree stump, broke a track. We were stuck at an angle with the underbelly facing the enemy and no means of depressing the gun to fire. Lt. 'Dickie' Payne then came by and, stopping, instructed me and my crew to abandon my tank and climb onto the back of his tank. Under cover of smoke we made our way back to the Regiment with just four tanks out of an original nineteen.[3]

With the motor platoons up ahead, Captain May had to deploy Company HQ as flank protection in case the *Hohenstaufen* followed up with infantry. Following this disastrous encounter, matters were becoming serious. General Roberts, however, focused on taking the battle to the enemy and capturing Vier. He wrote that: 'it might be argued, enormous chaos can be inflicted on the enemy by an armoured division thrashing around in his rear areas.' Clearly there was a probability that 11th Armoured Division and the 9th (*Hohenstaufen*) SS Panzer Division were set for a serious encounter battle but General Roberts 'decided that we should hold these two very dominating ridges [Point 218 and around Le Hamelière above Chênedollé] and then the Germans can attack us. Since we would be within the range of the corps artillery, we could inflict very heavy casualties.'

The advance of the leading company squadron group continued and when they arrived at Chênedollé at 1700 hours, the village was not held by the enemy. The tanks of B Squadron drove on past half a mile to the high ground between the village and the main road (RUGBY). Meanwhile, back in the village, as recorded in 8 RB's war diary: 'H Coy [and a troop of tanks] patrolled forward into CHENEDULLES [sic] and found it empty, but at 1800 hrs some tanks [again the Tigers of *Obersturmführer* Kall's No. 1 Company] and inf of 9 SS Pz Div appeared. H Coy got direct hit on a panzer with a PIAT but were later withdrawn.' More detail is added by Corporal Ault:

> After a slight skirmish at the top of the hill, the vanguard moved down into CHENEDOLLE, where tank opposition was met, causing the Carriers and Motor Pls. to play a most unpleasant game of hide and seek. One tank was effectively dealt with by Sgt Triggs, using the PIAT, and knocking out all the crew except the driver, who managed to reverse down the road.

Having earlier been happy to hold ground as far south as Chênedollé, it was rapidly becoming obvious that extending across the road was unwise and the battlegroups were ordered to halt, taking up two 'box positions' per brigade. Further back, the gun positions of 13 RHA and 151 Field also adopted local defence measures. Having been halted, shortly afterwards 8 RB and 23 Hussars

The PIAT was the British infantry platoon's anti-tank weapon, equivalent to the US Bazooka and the German *Panzerfaust*. All three fired shaped charge warheads but the PIAT, unlike the other two rocket-propelled weapons, was propelled by a heavy spring.

were ordered to withdraw a mile to the reverse slope of the Perrier Ridge below the hamlet of Le Bas Perrier. This position, occupied by 2000 hours, was less exposed and within a more comfortable range of the AGRA's guns but was still the position at the tip of the 11th Armoured Division's salient, which was under attack by the *Hohenstaufen* all the way back to Saint-Charles-de-Percy. Corporal Ault noted that:

> The company were withdrawn from the village while it was shelled by 25 pdrs and mediums, and then pulled back still further as the Guards Armoured Div on the left flank had been held up. The night was spent in close leaguer with 23 Hussars at LE BAS PERIER.

G Company recorded that: 'This high ground was extremely vital to the Germans, and we confidently expected to be heavily counter-attacked at first light next morning. Digging up here was very difficult but our perseverance paid ample dividends, as will be seen by the events of the ensuing days.' In the new position, H Company, which had been leading, were deployed in depth behind the other two companies and began preparing their positions. Sergeant Hicks was looking for a covered position in which to site his 3-inch mortars:

> I found a smallish recess cut into the rock face just off the road. It had probably been quarried out at sometime but it was an ideal place in which to site our two mortars to protect the rear of our position. The two carriers were conveniently parked nearby and a selection of ammunition was off-loaded and stacked in a corner against the vertical rock wall which rose about 20-odd feet above our head. The mortars were erected and sighted on to the valley below us. Food was cooked on the small petrol cookers and eaten. The tank battle had died away with our withdrawal and the barrage from our own guns eased off and stopped as darkness fell. We settled ourselves down in our mini quarry, wrapped ourselves in a couple of blankets on the rock floor and slept as best we could.

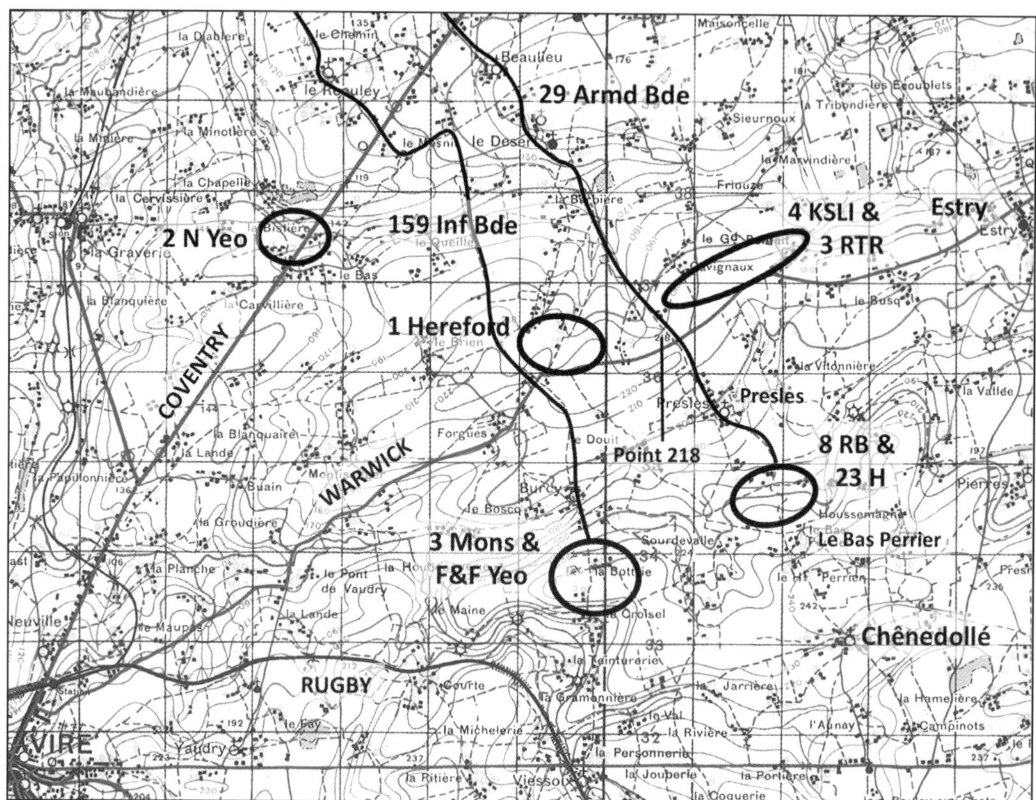

The 'box' positions taken up by the 11th Armoured Division's battlegroups during the evening of 2 August 1944.

Cut Off and Surrounded

The view of the ground that was to become 8 RB's most testing battle was revealed as the sun rose on the morning of 3 August and is well described by the 23 Hussars' historian:

> Our position was on a hill of the usual Normandy type, covered in high hedgerows, banks and cornfields. It was not big enough for the number of tanks we had, and overcrowding could not be avoided. The road from Presles to Chenedolle ran over this hill and it divided 'B' Squadron, who were on the west side, from 'C', who were responsible for the east. 'A' Squadron were in reserve on the northern slopes of the ridge and the Rifle Brigade companies were, of course, still with us.
>
> Our hill was a high one, and a splendid panorama of unspoilt Normandy country lay behind us. Directly to our north the ridge of Point 218, running parallel to our own, was visible two miles away, and beneath it lay the little village of Presles, which we had liberated the day before. Between us and Presles stretched a rich green valley, running east and west, apparently open, but in reality honeycombed with sunken lanes, banks, high hedgerows, orchards, and all the other intricacies of the 'bocage', terminating in thick woodland to the east. Far in the eastern distance, we could see the spire of

The view north from the area of H Company's and A Squadron's positions as described by the Hussars' historian.

German-held Estry, the rock which was later to break the assault of two successive British divisions and which was the last fortress to be abandoned by the Germans when they began to disappear into the vortex of the Falaise pocket. But although we could see clearly to the north, the view to the south was very different. Below us, completely obscured by orchards and woods though only half-a-mile away, was Le Bas Perrier. A wooded ridge immediately south of our own prevented us from seeing Chenedolle, but it also looked as though it would be impossible for many Germans tanks to engage us from it at the same time, owing to its undergrowth. The trees, however, stretched down into the little valley to within 200 yards of our position,

The much shorter field of fire south to Le Bas Perrier and the approximate location of the 'forward' motor companies.

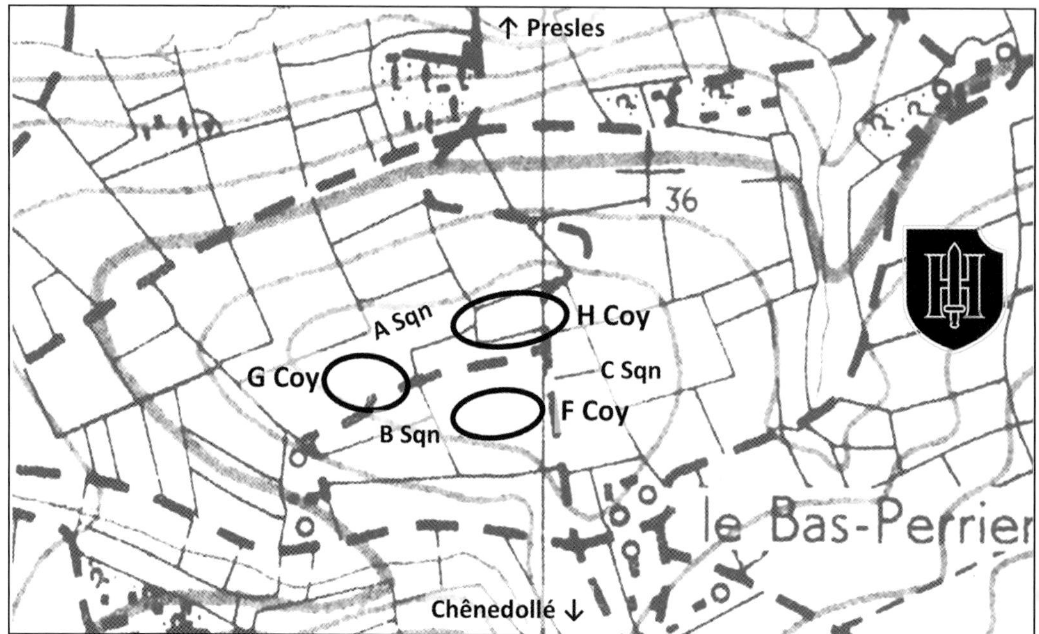

The deployment of the 8 RB and 23 Hussars battlegroup on the hill north of Le Bas Perrier. No grid references are given in the war diaries but with sketch maps from histories and ground observation these are approximate positions.

making a covered approach easy for an infantry assault upon us. Knowing that Chenedolle had been held the night before, we expected trouble to come from the south, but a keen watch was kept all round our hill.

Having dug trenches and cleared fields of fire for much of the night, Major Bell of G Company wrote that:

> The expected dawn counter-attack, to our surprise, it did not materialise, but it was not long, however, before we realised that we were in for a warm day. German infantry could be seen streaming back into Presles, thus cutting our centre line, and shortly after this, tanks and self-propelled guns, most concealed in sunken roads, opened up from all sides. 'Moaning Minnies' added to the fun and soon tanks began 'brewing up' and casualties began to mount.

The battle on the La Bas Perrier hill developed during the course of the morning from the arrival of enemy infantry into a tank v tank battle, to which the riflemen, through bombarded by mortars, artillery and *nebelwerfers*, were largely onlookers. From this point onwards 8 RB's casualties mounted.

H Company, having fought for much of the previous day, were looking forward to the other motor companies taking their turn bearing the brunt of the action:

> The next morning, we woke up to find that our position of the night before had changed considerably, and whereas our Company had been put at the rear of the Battalion position to have a quiet time, in the morning

some German tanks had come round to the rear into PRESLES and had succeeded in cutting us off. This was unfortunate, as we were rather concentrated and the vehicles must have made sitting targets because they destroyed several vehicles quite early on in the day, and the hill was a mass of burning hulks.

Before the battlegroup was properly cut off on the hill just north of Le Bas Perrier, a troop of self-propelled guns from 119 Anti-Tank Battery slipped through. A little later, an attempt by the A1 Echelons to resupply the battlegroup ran into trouble. The Germans had by now closed in on Presles in force, and H Company watched 'transfixed with horror', unable to help, as the logistic vehicles of A1 Echelon drove towards the village.[4] The vehicles, including the REME fitter's half-tracks and ambulances, were shot up by a pair of panzers in Presles. Some men were taken prisoner and others were wounded, while a few escaped. One RAMC private attached to the battlegroup would not take cover and went to look after a seriously wounded soldier:

> He refused to obey the Germans' orders to leave him and, despite the numbers of the enemy troops which surrounded him by this time, he calmly walked past them with his patient, and eventually regained friendly territory. Whenever a German stopped him, he pointed to his Red Cross armlet, and on every occasion they let him pass. It was a remarkable display of courage and endurance.

A M10 Tank destroyer of 119 Battery, 75 Anti-Tank Regiment.

These were soldiers of 9th SS Panzer Division; it is unlikely that soldiers from either the *Hitlerjugend* or *Leibstandarte* would have respected such an act of brazen courage, even by a medic.

Also reaching the battlegroup was Sergeant Jackson's Firefly, whose gun had been repaired by the REME fitters at A1 Echelon. Radio operator Trooper Slarks recalled that as they passed through Presles 'the thunder of the guns grew louder' and:

> At last we approached the battle area and drove off the lane to ascend the hill to join our weary mates of C Squadron. The noise was deafening as a Tiger tank armed with the deadly 88-millimetre gun picked off our tanks from its camouflaged position. All hell was let loose as our crews fought to defend our little hill. Over the wireless came, 'Who's that bloody tank coming up on the right?' In the heat of the battle I remember blurting out, 'Sergeant Jackson's tank'. And then we were in the thick of it! We managed to repel the attacks and after what seemed an age the enemy pulled back …[5]

A war diary entry by one of the staff officers of 29 Armoured Brigade summed up the enemy's tactics along the division's left flank:

> Enemy tks generally singly supported by a pl of inf, started infiltrating into our posns and on several occasions during the day the CL was reported as cut. In the very close country and without sufficient inf to piquet all the EAST - WEST approaches it was impossible to prevent this or to estimate accordingly the str of the enemy force involved. A more serious threat appeared when the enemy reoccupied PRESLES 7135 cutting off 23 H and 8 RB. No fwd mov was possible by any units.[6]

The division's left flank was under pressure from the *Hohenstaufen*, the 8 RB and 23 Hussars' battlegroup in particular. Overnight, the 3rd Division had come under command of VIII Corps and 185 Brigade was put under command of General Roberts at 1100 hours as a reinforcement. It was originally intended to insert these battalions into the line to bolster the two brigade positions but owing to the situation on the left flank, 185 Brigade had to be deployed in order to protect the division's supply lines from infiltration and being completely cut off.

Up on the hill, the fighting continued unabated. The panzers would typically appear in gaps between the houses of Le Bas Perrier and Presles or other cover, 'quickly sight on to a target, fire at it, and immediately take cover again'. With Germans having limited options and not always able to benefit from their superior range, in a protracted battle, the battalion's 6-pounder anti-tank guns, the tank destroyers and the Hussars were able to anticipate where the enemy would appear and managed to knock out a number of panzers, ensuring that it was not a one-sided battle.

To add to the confusion, as noted by 13 RHA gunner officer Lieutenant Delaforce, 'American Thunderbolts were notoriously trigger happy. They swarmed over the thin red line of British defences and bombed everyone despite

One of 13 RHA's 25-pounder Sextons, built on the RAM chassis.

the fluorescent orange panels and the orange smoke emitted from smoke generators.' In another incident, British Typhoons circled the hill and dived to attack the riflemen:

I was looking at the leading aircraft straight in the eye so to speak, realised he meant business and promptly dived headfirst into my slit trench and made a very uncomfortable landing. The leading pilot must have seen the yellow panels and smoke canisters simultaneously with him pressing the firing button and called off the strike as the following pilots didn't fire. They circled around a few times quite low and by this time all our positions were given away to the Germans by a mass showing of yellow panels like an erupting daffodil field, but the Typhoons realised they had made a mistake and flew off. Unfortunately, the one salvo of rockets from the leading plane was dead on target, blowing the turret and gun completely off the Sherman tank of the 23rd Hussars, which was by a hedge about 25 yards from my slit trench, killing all the crew.

We learned later that the pilots had identified the wrong hill but seeing all the shell bursts all around it thought that we must have been Germans fighting to hold on to the hill instead of course the other way round.

Sergeant Hicks, in the cover of his small quarry on the rear of the hill, recorded his feelings as the morning progressed:

> Finding ourselves cut off from the rest of the Brigade group was not a very pleasant feeling to experience. We felt more than a little naked as if we had been caught with our pants down and in addition our vehicles were much too closely grouped together on the side of the hill and presented a very vulnerable target as not all vehicles could find adequate cover or be suitably camouflaged.
>
> We soon became aware of our vulnerability as both Tiger and Panther tanks in the valley began to pick out their targets on the hillside with some accuracy. One of our support Churchill tanks [one of six AVREs], one of the largest and heaviest tanks, was parked on the road barely 20 yards from our position. It must have been either visible through a gap in the hedge or silhouetted against a lighter background, but whatever the reason an 88mm armour-piercing shell hit it, causing serious casualties and completely wrecking it.

The number of casualties from the increasingly heavy enemy tank, artillery and *nebelwerfer* fire grew steadily and they could not be evacuated. This left the Regimental Aid Posts of both 8 RB and 23 Hussars with more and more wounded to look after. The Padre and captains Wilcocks and Mitchell, the two regimental medical officers, along with their RAMC orderlies and stretcher bearers, received universal praise for their stalwart work during the time the two units were cut off at Le Bas Perrier.

A medium gun of 8 AGRA in action in Normandy. The part played by the artillery in preventing 8 RB being overwhelmed at Le Bas Perrier is difficult to overstate.

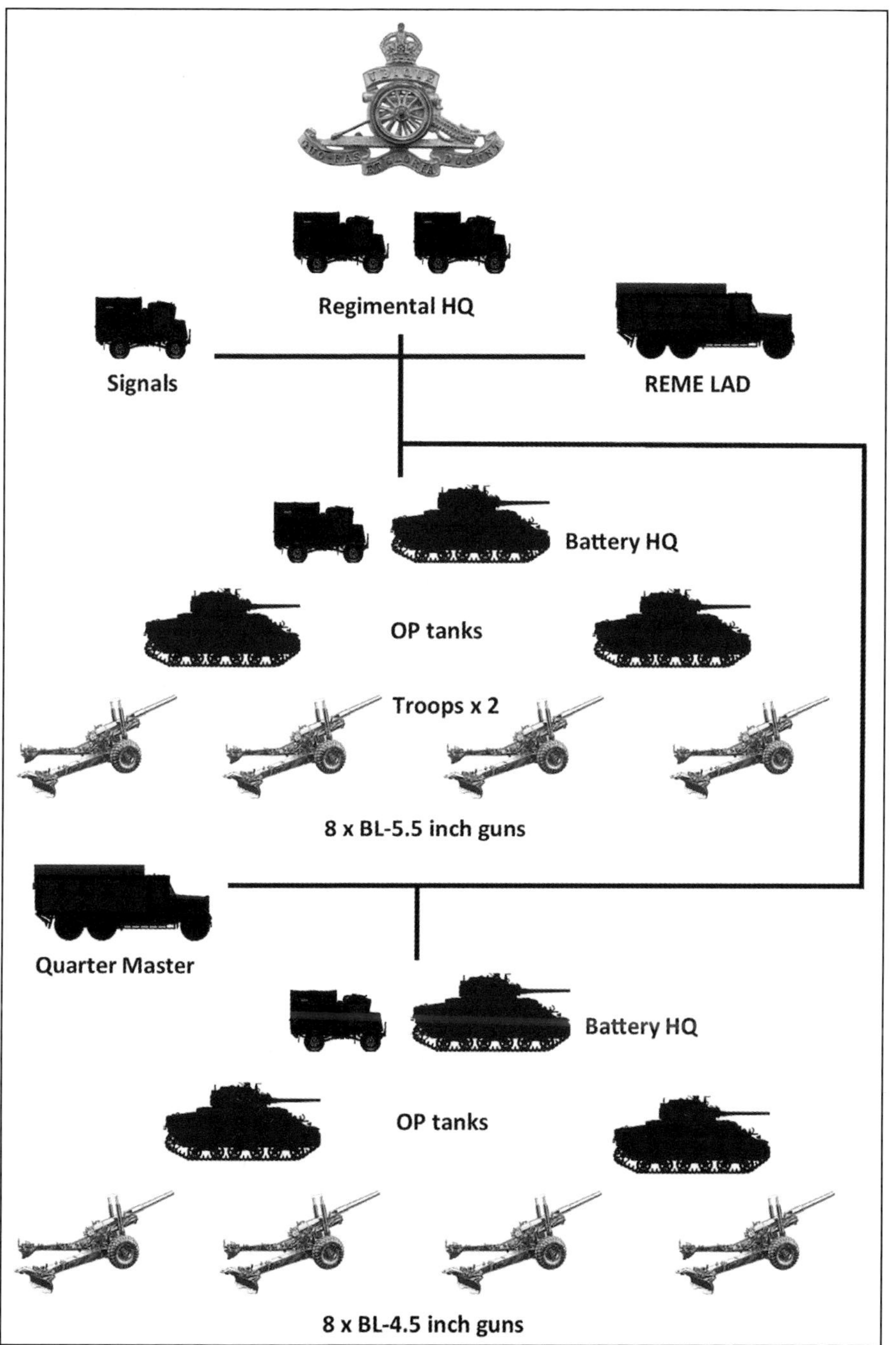

The organisation of a medium regiment, including armoured observation post tanks (usually provided by the supported formation), as found in 5 Army Group's Royal Artillery that supported VIII Corps throughout the Normandy Campaign.

As the battle on the hill continued, the rifleman on various parts of the defences were able to take compass bearings of the launch signature of the *nebelwerfers* that were firing salvo after salvo at the battlegroup. The grid reference of the resulting intersection was passed to one of the medium regiment's OPs and the guns of the AGRA produced a satisfactory firework display at the target end as presumably an ammunition dump was hit. General Roberts, who normally spent much of his time away from the headquarters visiting, explained that:

> During this time, I personally stayed at Divisional HQ. This had the advantage that I had a telephone line to Corps HQ and when a German attack started building up, I could talk at once, not only to the BGS (Harry Floyd), but also to the CCRA. At moments of crisis, and there were one or two, I asked the CCRA for absolutely all he could lay his hands on and the support really was terrific.[7]

Sergeant Hicks had established an OP near H Company's position 'well hidden in a ditch and bank of a hedgerow from where we could command a reasonable view of the houses and road running along the bottom of the valley'. During the afternoon he explained that:

> there was an almighty explosion behind us and we saw a cloud of smoke and dust rising up in the vicinity of the mortars. I tried to get Jack Nixon on the radio to find out just how close that one was, and it seemed ages before he replied in a very shaky voice that they had received a direct hit. We abandoned the OP and raced back to find an horrendous scene. Jack was lying wounded by his radio, Phil and Spencer were also wounded and wandering about very dazed. Dusty Miller and Nobby Littlepage were both still sitting on their jerrycans leaning against the wall just as we had left them but both of them had the tops of their heads sliced off at eye level, just like the top taken off a boiled egg. I was completely horrified and devastated. Jack Bachelor, who had been sitting between them, lay on the ground stunned but miraculously unhurt.
>
> We patched up the wounded as best we could and with the help of other lads we got them through to a field about 100 yards or so across the hill where the MO (Butch Wilcox) had set up his field dressing station alongside a big hedgerow with the MO of the 23rd Hussars.

The seriousness of the situation on the hill at Le Bas Perrier was realised by brigade and division and, as recorded in 29 Armoured Brigade's war diary, at 1200 hrs '2 WARWICKS from 185 Inf Bde placed under command to clear CL on PRESLES by night attack'. This they did, as described in their war diary. 'The Bn attacked at midnight, supported by 11 Armd Div Arty, with "C" Coy on the right and "D" Coy on the left, followed up by "A" and "B" Coys.' They were fully supported by artillery and:

> The attack was successful [the enemy had moved on] and the Bn consolidated in the village of PRESLES. No casualties in the attack and by the

early morning the enemy began to shell the village and some casualties occurred.

Through the early morning fog of 4 August, H Company dispatched patrols down to Presles and successfully made contact with the Warwicks and fifteen ambulances were soon on their way up the hill to evacuate the wounded.[8]

Most of the German armour had moved on towards Vier, but regular heavy shelling and mortaring made the day just as bad, with the *Hohenstaufen*'s panzer grenadiers resorting to attempts to infiltrate forward onto the hill. In response, the divisional artillery fired regular concentrations around the battlegroup's positions to deter approaches along ditches and hedges. 8 RB were, however, able to mount tank-hunting patrols, one of which was from G Company:

The 4th of August was another day of being shelled and 'minnied' and a 'PIAT gang' went to Chenedolle. It was, however, much more comforting than the previous day inasmuch as we knew that at least we could not be fired on from our immediate rear. Our padre, Jeff Taylor, who had been up with us the whole time, held a simple impressive burial service for those killed on the previous day, during which service, perhaps by mere coincidence, not a shot was fired by either side.

Later in the day, the Warwicks' war diary noted that: 'At 1300 hours orders were received from 11 Armd Div to take over BAS LE PERRIER from the 29 Armd Bde's Motor Bn, 8 Rifle Brigade, who were, at the same time, to come back to

A Typhoon of 2 TAF being scrambled at an Advanced Landing Ground somewhere in Normandy.

PRESSLES. This change over was to be completed by 1800 hours.' G Company recorded that:

> Their reconnaissance parties arrived up in the morning and were heralded by the arrival of some misguided Typhoons who, having fired one rocket into our area, were dissuaded from further aggressive action by dense clouds of yellow smoke – our recognition signal for friendly aircraft.

According to 8 RB, later the 'Relief was interrupted by an enemy attack of 2 Coys Inf sp. by tks, which we [and the tanks] drove off together'. Heralded by 'ten very unpleasant minutes of nebelwerfer fire', this attack was launched from the woods near Le Bas Perrier, as recalled by Sergeant Hicks:

> … the German infantry defied all the modern theory about keeping well-spaced apart and came forward in a mass almost shoulder to shoulder. Our 25-pounders laid on a non-stop barrage on to them as well as all the small arms fire our leading company (which I think was 'G' Company) poured into them as well as our tanks. Ammunition was getting short and as my mortars were not being used I received orders to get all my mortar bombs forward to 'G' Company mortars.

In the midst of a relief in place is possibly the worst possible moment to be attacked and the Hussars observed that: 'Colonel Hunter of the RBs rallied his mixed and rather disorganised force of infantry to meet the attack.' Mortar fire into the treetops greatly helped beat the enemy back and the battalion's war diary concluded that the enemy paid a high price for a type of attack favoured by the SS on the Eastern Front. 'It appeared later that the hy cas suffered by the enemy during this attack were locally of decisive importance. The relief took from 1800 – 2200 hrs, vehicles being left behind and brought back after dark.'

Presles

For the riflemen, dawn on 4 August revealed that the village was 'a very different place to the PRESLES we had liberated two days before, where the church bells had been rung in our honour and where there was more wine than even the thirsty Riflemen could drink'. The move back to Presles was, however, not to a rest camp, as the division was far too thinly spread for that, even with 185 Brigade under command.

The battalion occupied defences around the paddocks and orchards on the hillside immediately north of the village, again with a view to all-round defence. Even though the distance between battlegroups up on Point 218 and Le Bas Perrier was reduced the threat of enemy infiltration through the divisional area remained. The whole process was aided by the deep trenches the Warwicks had dug during the day, which were more numerous than needed by the by now sadly depleted 8 RB. Due to the number of the battalion's vehicles knocked out or damaged in the centre of the defences on the hill, 'except for a few command post vehicles the remainder were concentrated in a hide a mile to the rear to minimize

Soldiers on sentry duty man their section's Bren gun.

the further loss from shelling'. The H Company history noted that: 'We were very spread out at PRESLES and the Company position stretched out along the forward slope going down into the village, covering in all an area about 500 yards long.'

During the evening of 5 August, 2 Warwicks with a more numerous four-company ORBAT being properly dug in on the Le Bas Perrier hill and enemy activity reduced, two squadrons of 23 Hussars came back through the riflemen's positions at Presles, to refit south of Le Désert. Being primarily defensive weapons, a troop each from the divisional and corps anti-tank regiments, both SP and towed guns, replaced the Hussars' remaining squadron the following day.

Although shelled for the five days spent in their trenches above Presles, the enemy confined his offensive efforts to Le Bas Perrier, making only one attempt at infiltration of 8 RB's positions that was broken up by artillery fire. Major Bell later wrote:

For five more days Presles was to be our home, a period spent for the most part in our extremely near slit trenches and clad in our steel helmets. These precautions were very well worthwhile, and considering the intensity of shelling and mortaring our total of a dozen casualties was not unduly high.

The nights on the whole were quieter than the day, and there was most mornings a very welcome early morning mist which, while it lasted, enabled us to move about quite freely. One night, however, the 'Minnies' came down in no uncertain manner, causing havoc in the orchards and setting hedges and trees on fire, as well as a house in 10 Platoon's area full of teller mines and explosives.

With good deep trenches that the companies covered over with logs, planks and anything that would protect them from airburst shelling and mortar fire, life was

The deployment of 29 Armoured Brigade in the Presles area from 4 August.

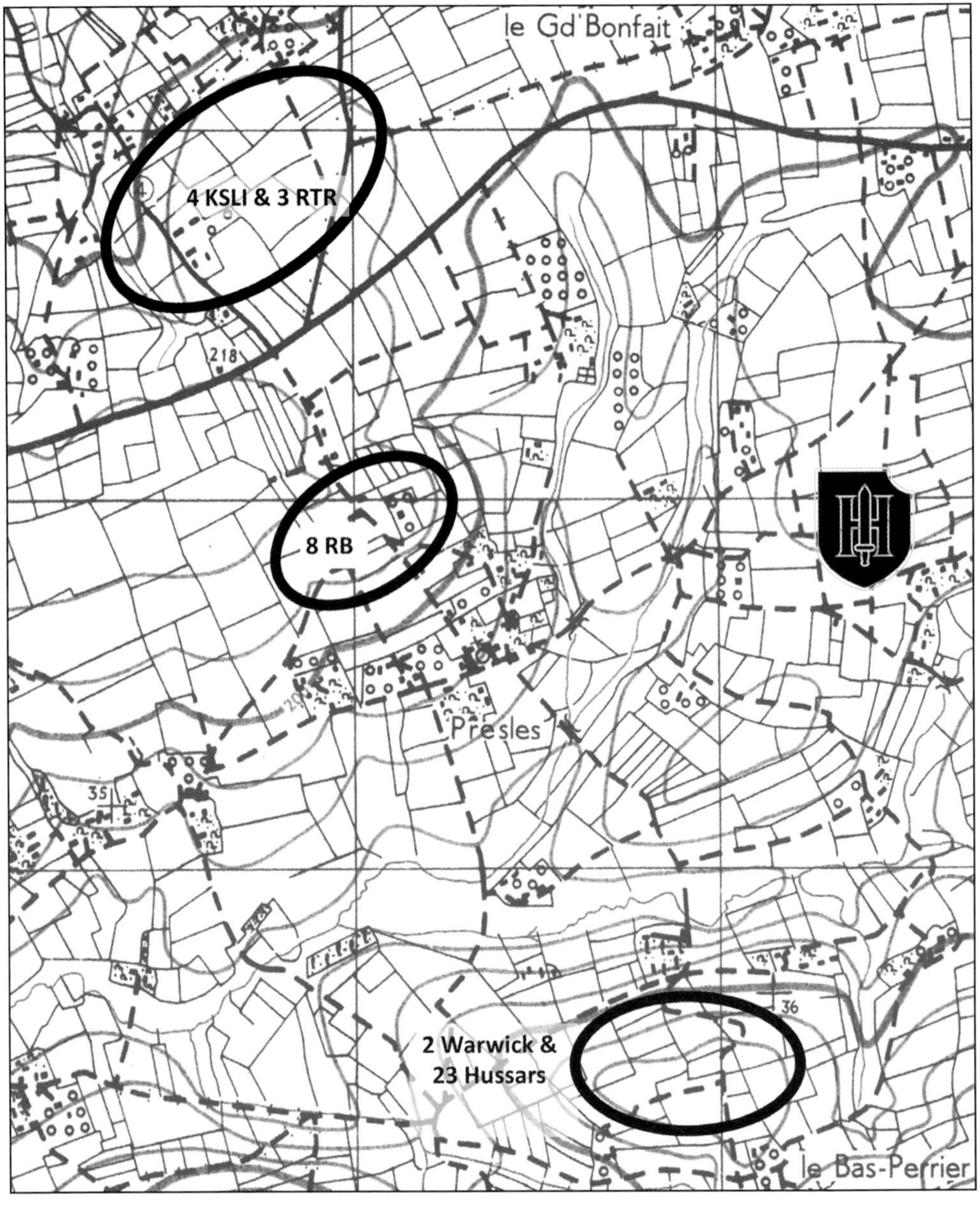

largely subterranean. The riflemen 'didn't venture far from our dugouts but after a couple of days we learned by the sound of incoming shells whether it was going to be a close one or not and when we could expect there to be an interval to dash around and collect rations etc. and of course to answer the call of nature.'

For those who had access to radios, tuning into the BBC news was one of the few ways of keeping in touch with the world beyond the battalion. On 4 August the riflemen listening to the Evening News were outraged to hear during a piece on the breakout that the Guards Armoured Division were 'the spearhead of this party'. The riflemen all knew that the Guards had begun their part in the operation days after them and were still miles to the rear.

By 10 August the Second Army was now beginning to advance again and the Germans showing signs of withdrawing from the Sector west of the Orne. The 43rd Wessex Division had captured the dominating Mont Pinçon, the Mortain counter-attack was being defeated, and the US Armies were spilling out into central France and Brittany. To be ready for the next phase of the campaign, 29 Armoured Brigade was withdrawn to reorganise and refit. After days in contact, the prospect of rest for the battalion was alluring but:

> Twice we were going to be relieved from this unhealthy position and twice for various operational reasons we were disappointed at the last moment. Eventually on the night of 10th of August the magic code word 'SWOLLOW' came over the air, and with a sigh of relief we streamed out of our positions back to our transport, which carried us back to the Beny Bocage area again, where the Echelon had a most welcome hot meal awaiting us, and where we needed no encouragement to lie down and sleep.

As noted by F Company:

> The Battalion was eventually relieved by the 5th Battalion Coldstream Guards, and we withdrew to peace and quiet in the shady fields round Le Beny Bocage. This was the end of a period of fighting which had lasted about two months, and which was never approached in intensity or bloodiness during the remaining nine months of the campaign.

General Roberts, in a special order to all ranks, wrote summing up the period up to 12 August:

> It may be of interest, but perhaps not of comfort, to realise that the greatest enemy resistance is at the present time on 8 Corps' front. By containing the enemy's strength we are being the utmost value to the remainder of the army.

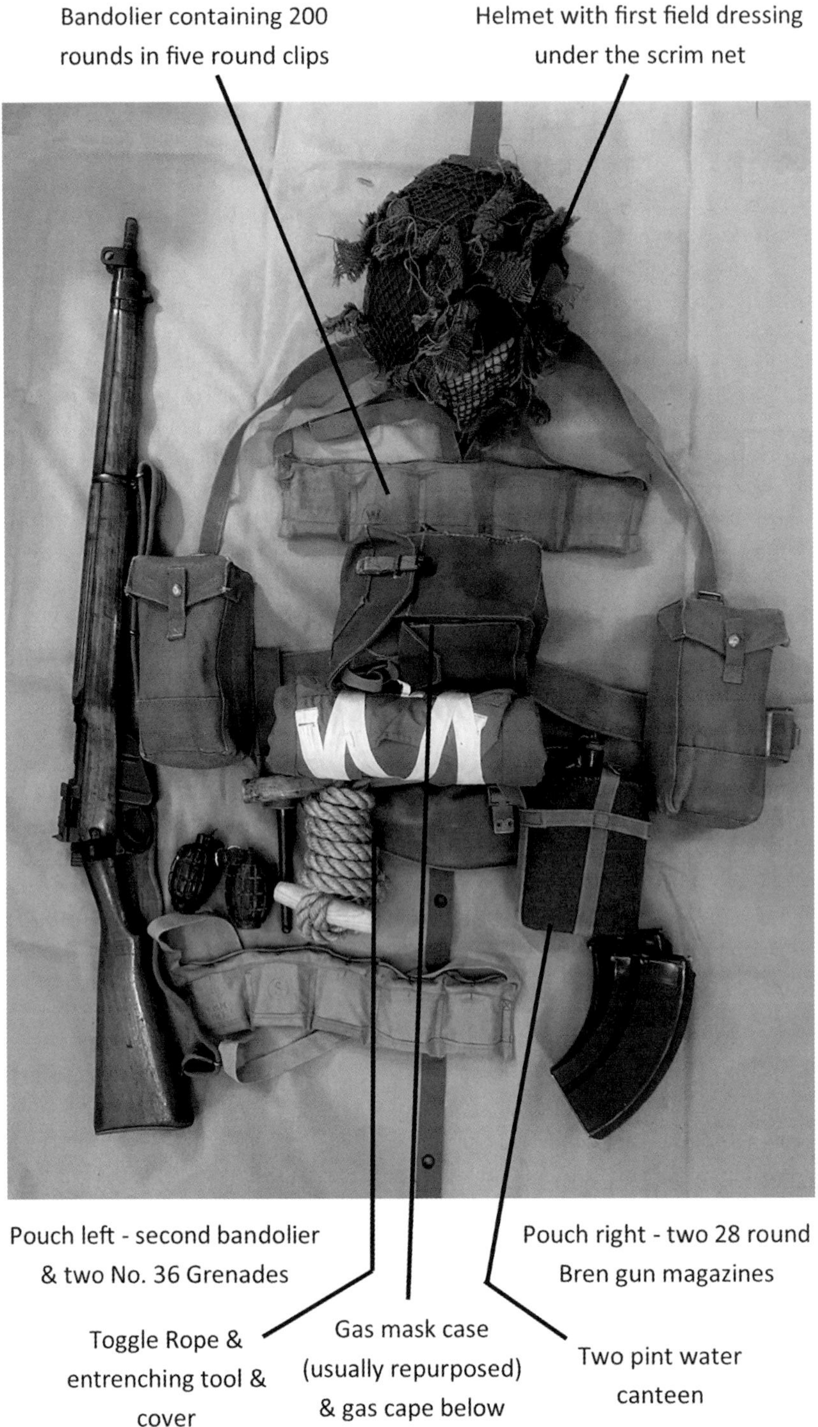

The rifleman's basic 37 Pattern web equipment and its contents. (*Matthew's Military Moments*)

Chapter Eleven

Pursuit to the Seine

Following the EPSOM, GOODWOOD and BLUECOAT battles, 8 RB was some 250 riflemen below strength, which was particularly felt in the motor platoons, as in any mechanised or armoured unit finding drivers and vehicle commanders is always a priority. While at Le Bény-Bocage, a whole company of riflemen, numbering 170 strong, from 8th King's Royal Rifle Corps (8 KRRC), joined the battalion. They had been a part of the motor battalion in 9th Armoured Division, but such were the rates of casualties in Normandy, particularly among the infantry, this division was disbanded on 31 July 1944, with whole units, sub-units and individuals becoming reinforcements.[1]

As 8 KRRC riflemen were motor battalion trained, to help them settle into their new battalion, as much of their structure was maintained as practical. Each motor company received a platoon and a scout section. Major Bedford took over command of G Company. Despite this reinforcement and a trickle of men returning to duty from hospital/rehabilitation, the battalion started the next phase of the campaign still fifty-seven men under strength.

By the second week in August, with the Germans having held on defending ground, at Hitler's insistence, for too long, the potential for the envelopment of a large proportion of the German armies in Normandy was presenting itself. The First US Army, having broken out from Normandy, was swinging east, the First Canadian Army battling south towards Falaise and the Second Army was squeezing the developing pocket from the north-west, with XII Corps on the left and XXX Corps on the right. As VIII Corps was being squeezed out by the converging advance, in the next phase, during Operation GROUSE, 11th Armoured Division was to come under XXX Corps, now commanded by Lieutenant General Horrocks.

General *Feldmarschall* von Kluge asked for and finally received Hitler's permission to withdraw behind the River Seine. He ordered the remnants of his panzer divisions back from their abortive counter-attack at Mortain and had ordered the infantry divisions facing the Second Army to stand. Once the panzers that could be saved were out of immediate danger, an orderly withdrawal was to be conducted and the first signs of the German withdrawal were spotted within days but were treated with some suspicion by the riflemen, as recalled by Major Bell:

> ... we still had fresh in our minds memories of earlier battles, where each time we were told the enemy was pulling out, a violent counter-attack developed and we resolved to decide for ourselves when the withdrawal could be publicly announced.

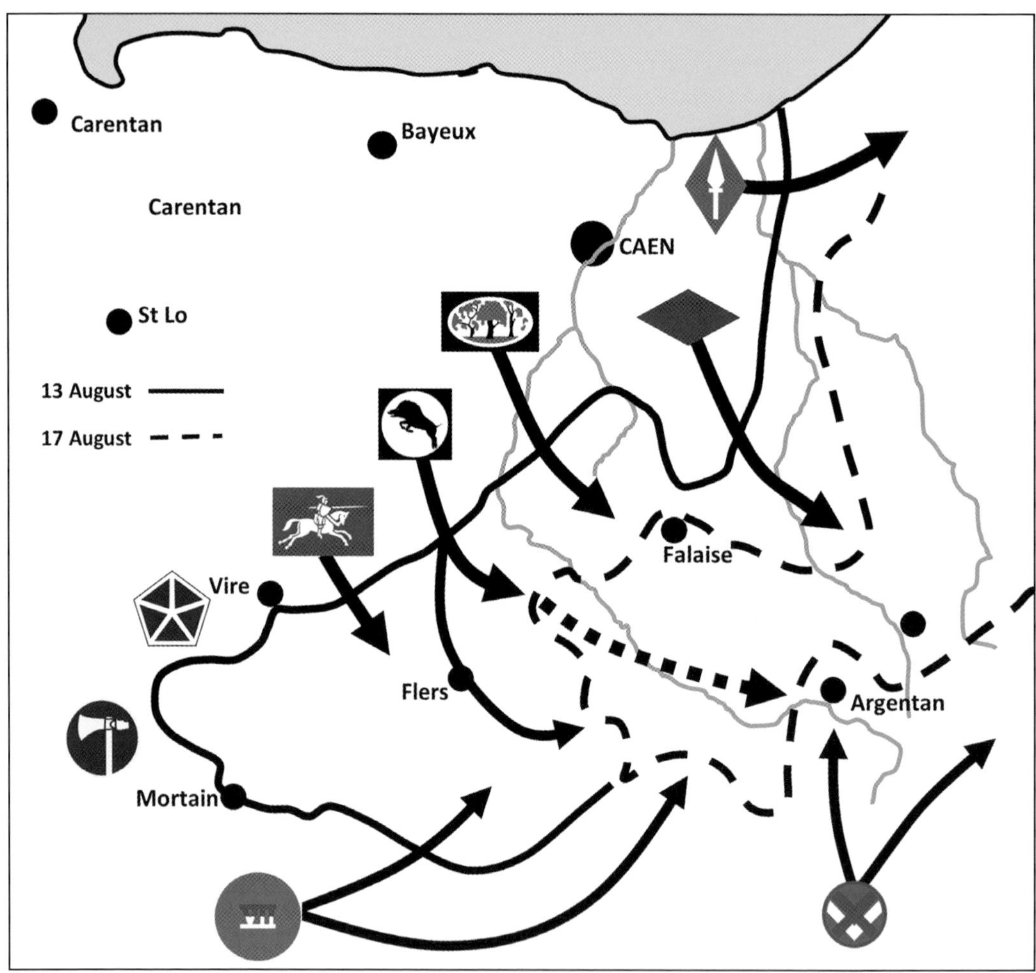

The envelopment and 'squeezing the pocket' by the Allied corps.

Operation GROUSE

During the morning of 12 August, after less than forty-eight hours out of the line, 8 RB received a warning order to move that night, with the operation order following at 1600 hours confirming that 29 Armoured Brigade was to take over from 46 (Highland) Brigade of 15th Scottish Division to the north-east of Estry. The task given to 8 RB was 'with under command 13 Fd Sqn RE, [8 RB] will relieve 9 Cameronians, and one Coy of 2nd Seaforths in the area DRUERIE 7641, and will move in accordance with the march table attached'. On the limited routes available the brigade began its move to relieve the Scots at 1730 hours but as the penultimate unit in the brigade's column, it was not until 2200 hours that the battalion was on the road. The newly arrived OC of G Company, Major Bedford, recalled that: 'The move up was very chaotic, chiefly owing to traffic diversions,' that continued:

> During the morning of the 13th, the day before the new offensive was due to start, General Horrocks visited us and explained the plan. At first we were

to advance circumspectly, clearing as we went, but as soon as the enemy appeared really to be on the run, then were we to thrust forward boldly, leaving at our discretion any hostile force less than a company strong to be mopped up by the 50th behind. This acceleration was not expected to be possible for the first few days and the 50th would not be available before the 18th. For the time being, therefore, the old method of advance continued, the two brigade groups, constituted as before, moving concomitantly and using as many roads as the sector afforded; for in the bocage you are confined to the roads and an armoured division has so very long a 'tail' that the more routes you can employ for the combatant troops the better … Places not traversed by the routes currently in use were normally investigated by armoured car patrols of the Inns of Court, who also searched the flanks and made contact with the formations thereon if there were any.

The corps' objective was the line Flers–Condé-sur-Noireau and the divisional objective was the town of Vassy. A daily account of the advance to the Seine can be assembled from unit and sub-unit histories, personal accounts and various war diaries.

13 August

The operation order issued during the day provides the detail of 29 Armoured Brigade's plan:

  8. The adv will take place in three PHASES.

PHASE I.

  9. Half an hour after first light [about 0600 hours] one armd sqn and one coy/motor coy will adv on each route to WHISKEY.
  10. When first objective is firm, one sqn and one coy/motor coy will be moved to pt 208, and to LASSY 7940 as a firm base.

PHASE II.

  11. On orders from this HQ, adv will be resumed to GIN.

PHASE III.

  12. Portion of bde remaining present posns area 7640 will be moved fwd to First Objective.
       Final layout will be: -
         Two sqns and two coys/motor coys on each final objective.
         Remainder of bde on first objective as firm base.
  13. It is essential that the first waves push on to the successive objectives and that any necessary mopping up is done by the second wave.[2]

The leading regiments, 23 Hussars and 8 RB on the right and the F&F Yeo and 3 Mons left, were to advance, with each squadron grouped with a motor or infantry company. Both battlegroups had in addition to their normal FOOs a troop of SP anti-tank guns from 119 Battery and a troop of engineers from

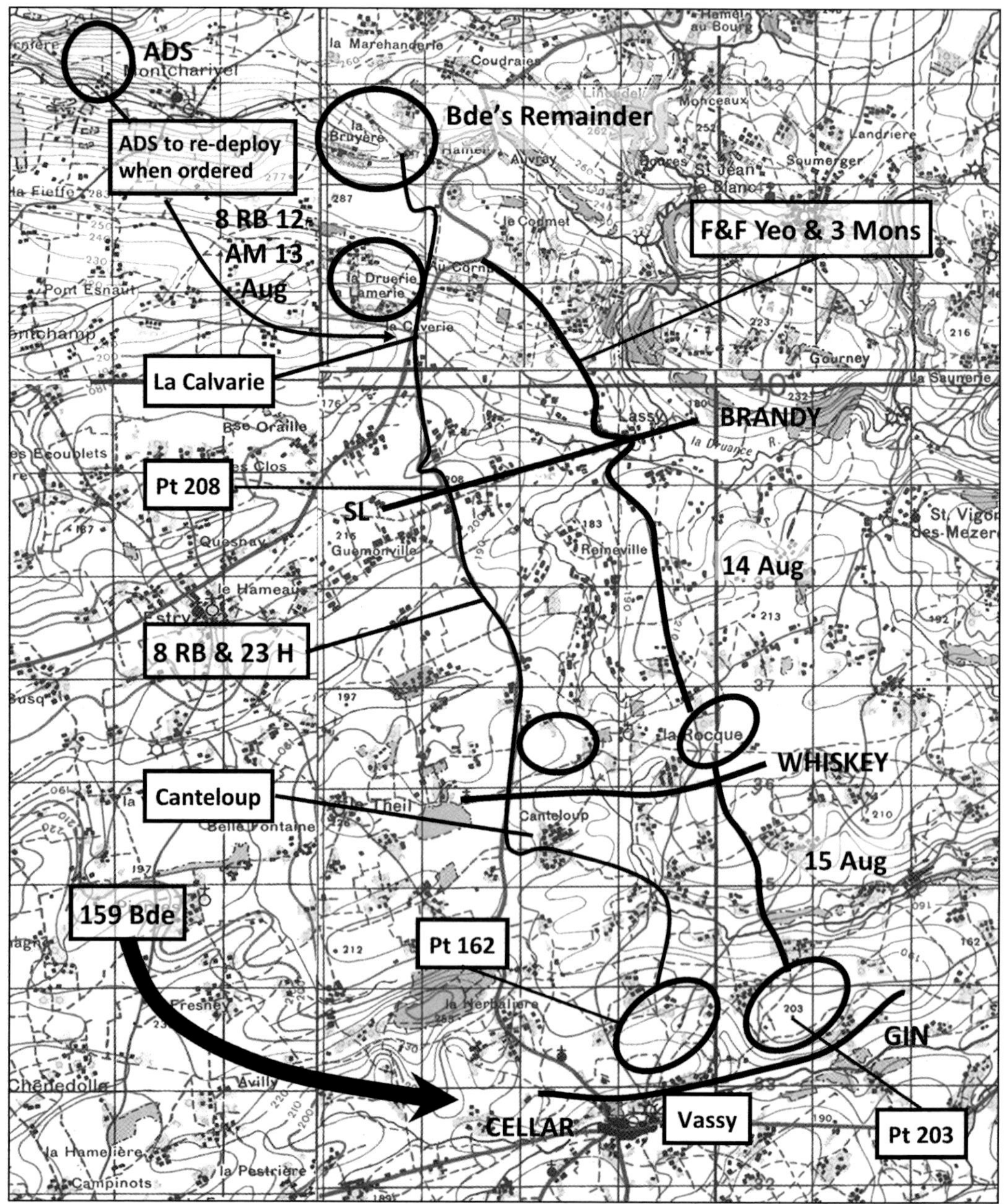

The 1:50,000 map marked up with details of 29 Armoured Brigade's Op O No. 10 showing their plan for Operation GROUSE, the advance on Vassy.

13 Field Squadron RE. Initially, Colonel Hunter's headquarters 8 RB was to move with Brigade Tac HQ.

With orders given and companies and squadrons physically regrouped, F Company, 8 RB, and B Squadron, 23 H, began their advance in the early afternoon and by 1700 hours the riflemen had secured Point 208 overlooking the following morning's start line. Blocking the road to BRANDY/start line, they encountered a blown-down tree and in the verges, mines. The Hussars' historian recalled:

> The road along which they were to move rose to a ridge of high ground [Point 208] which continued south for some miles until it dropped down into the Vassy valley. No sooner had they reached the northern edge of this when they were held up by mines and obstacles. The Sappers were called for but the clearing took longer than expected, and the Squadron withdrew for the night, leaving a troop in possession of an important cross-roads slightly in rear.[3]

Some of F Company's riflemen remained with the tank troop at La Calvarie, while the rest moved forward to Point 208.

14 August

Heading the advance, the armoured cars of C Squadron, Inns of Court Yeo, were delayed by half an hour due to fog, but so thick was the country they returned to La Calvarie, and by their own admission, 'no quantity of information was obtained'. Meanwhile, the sappers were at work clearing the remaining obstacles at first light, allowing B Squadron to join F Company on Point 208. Following immediately behind them was H Company and C Squadron group, who took up the lead, but as the leading tank approached WHISKEY, it was fired on and was knocked out by what was initially thought to be a pair of Jagdpanthers positioned astride the road. Major Bradford noted that: 'The whole Company deployed and 14 Platoon got very close to a Mk. IV. Unfortunately, they had not got their PIAT with them and so could do nothing about it.' The Hussars' historian adds:

> This combined manoeuvre finally established the identity of two German Mark IV tanks and forced them to retire. All this took the best part of a dusty, hot and sultry day. By four o'clock, however, the whole C Squadron group was established astride the ridge at its highest point, half-way to Vassy.

H Company noted that 'we eventually took up a defensive position on the reverse slope of a dominating hill [just short of WHISKEY]. G Company then passed through us late in the afternoon and continued the advance.' Major Bell described the riflemen's advance with A Squadron:

> After a quiet morning on the road leading south to Vassy, we passed through the leading group, and it was not long before the first tank was 'brewed up' outside the village of Canteloup, referred to over the radio as 'the fruity village,' which was found to be held by the enemy.

Nestling in a valley, Canteloup was largely obscured by foliage, and it was initially difficult to establish if it was occupied by the enemy, who held their fire until the Hussars' Shermans got just too close:

> An attack, supported by the tanks, was launched and 11 Platoon on the right pushed into the village without meeting serious opposition; 10 Platoon,

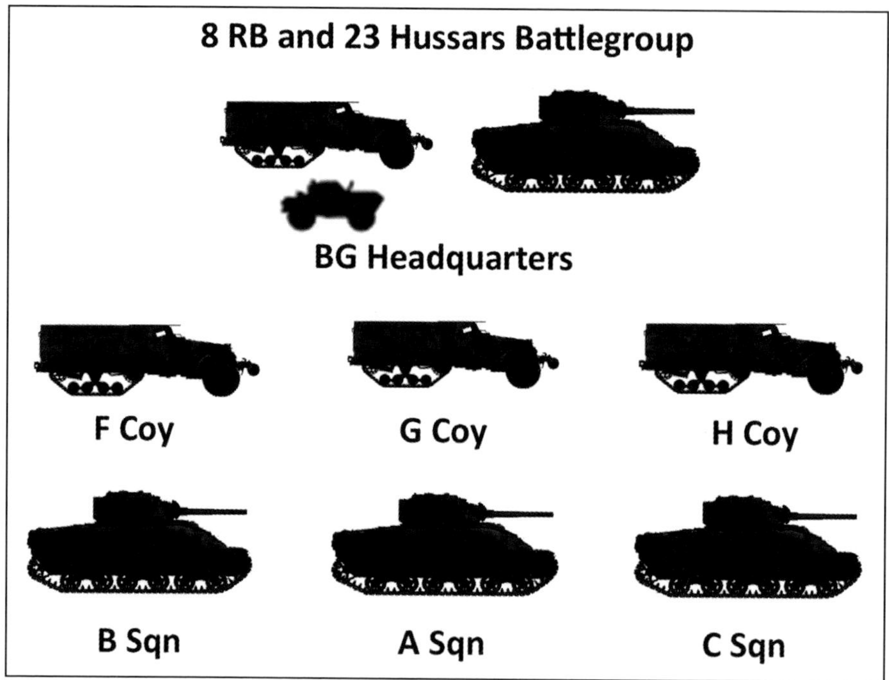

The grouping of 8 RB's motor companies and 23 Hussars' squadrons for Operation GROUSE.

however, on the left were held up and suffered casualties, three of them being fatal. With darkness coming on, it was decided to withdraw the motor platoons, leaving 9 [carrier] Platoon to consolidate on the cross-roads on the outskirts, while the remainder of us went firm farther back.

An amusing incident occurred at Company HQ during the attack, when a German appearing out of the hedge, and wanting to surrender, tapped Bishop on the shoulder, the latter being so taken aback that he nearly gave himself up to the German instead, and even his inevitable pipe was said to have fallen to the ground.

The brigade's diarist summed up the advance and noted that during the course of the day orders were modified:

Throughout the day the adv on both routes was slow, due to enemy rearguards sp by A tk guns and a few tks or SP guns and mines. The orders to units were that it is NOT the intention to push on as quickly as possible but to mop up every posn as they are discovered and put as much eqpt, etc. as possible in the bag.

Clearly the tone of the previous day's operation order had changed. Brigadier Harvey's clear intent to push on had been changed in the light of the difficulties of the ground and lurking enemy rearguards. Major Bradford recalled: 'That evening the Colonel arrived and the details of van-guards were worked out and went into operation the next day.' With difficult ground and a competent enemy rearguard action, the advance was clearly far more of an infantry enterprise than had been anticipated.

The advance on 14 August took the leading elements of the 8 RB/23 H battle-group 4 miles into enemy territory, just over half the distance to Vassy. 'A fine stonk that night,' is recorded to have 'destroyed a Mark IV tank and caused the enemy to abandon the village.' The enemy almost certainly already had planned to withdraw.

That evening at 2300 hours, to prevent Brigadier Harvey from pushing too far ahead of the rest of the corps and getting into another Le Bas Perrier-like battle, 'The GOC telephones and said "form for tomorrow – soft pedal – maintain contact follow up if enemy withdraws but units not to get involved."'

Infantry in the bocage, even motor battalions, spent a considerable portion of their time on foot.

15 August

The day started shortly after dawn with a:

> CO's conference HQ. Bde Comd explains GOC's message about 'soft pedal'. Early morning mist again prevented early resumption of adv but by 0800 hrs both columns were moving slowly, SL being blocked by demolitions and mines. Small but very few enemy rearguard parties encountered.

G Company and A Squadron resumed the advance, passing Cantaloup to the west, confirming that it had been abandoned by the enemy, and headed south to Point 162, 'a feature dominating the valley to the east of Vassy and the road to Conde'. Major Bell described the advance and arrival on the 162 feature:

> After an unpleasant night of shelling and mortaring, during which our anti-tank platoon from E Company suffered casualties, the advance in the morning continued without any serious opposition to the high ground north of Vassy, where we were sharply shelled on arrival. This subsided, however, and we settled down here for the night.

Due to the F&F Yeo battlegroup being mired in traffic, F Company and B Squadron were directed to secure Point 203 in order to have both of the brigade's objectives in the hand. As recorded by the 23 H: 'By a series of uncharted tracks B Squadron made their way through Aligny across to another hill [Point 203] more to the east and there spent the night.' The battalion's war diary for the period 12–15 August reads:

> The Bn. was at this time working with 23 H, Coys under comd of Sqns in open country, vice versa in close country. It appeared that the enemy was withdrawing slowly, so we pushed forward Coy and Sqn gps each day until 15 Aug we were just north of VASAY [*sic*].

During the advance, Sergeant Hicks' mortar section moved to successive base-plate positions, ready to answer calls for fire from the company. He wrote:

> I remember we had to pull off the road and deploy up a hill and on the top in a clearing beyond the trees were the signs of the battle in which one of our infantry regiments had been engaged. We had to stop there amongst a few burnt-out trucks and slit trenches that were dotted about the open ground. Not expecting to be there for very long, we hung about our vehicles and brewed up a cup of tea. The morning had been quiet with little or no activity being pointed in our direction. Quite suddenly we heard Moaning Minnies blasting off and screeching towards us. We all scattered and dived for cover, quite literally in my case as I went headfirst into one of the nearby slit trenches, which was a good one too, about four feet deep.
>
> When the shelling stopped, I stood up and peered around waiting for the smoke and dust to settle. It was then I realised the bottom of my trench felt spongy. At first I thought it was just loose earth that had fallen in. But

Scout and armoured cars of the corps' recce regiment, the Inns of Court Yeomanry, patrolled in front and to the flanks in order to identify enemy positions.

pressing with my foot in a couple of other places I became aware of that unmistakable odour and shuddered at the knowledge that I was standing on a corpse that had been hurriedly buried there with only a few shovels of earth to cover it. Curiosity almost made me investigate further and to find out if was one of ours, or one of theirs, but quite frankly I didn't have the guts to scrape away the earth with my hands to see the colour of the uniform it wore. I just got out and went back to my carrier and crew to see if all was well. My tummy felt a little bit queasy and I didn't mention to the others what I thought lay at the bottom of the trench.

When in the thick bocage, map reading with the 1:50,000 map proved to be extremely difficult. It had been copied from French maps and updated by air photography, with a common complaint being that some decent-looking roads on the map were in fact little more than tracks, some tracks marked on the map did not exist and yet more tracks as discovered were not marked at all. Reporting own and enemy locations could be a hazardous business, especially when calling for artillery and mortar fire. As noted by 23 H: 'All this was dull, dangerous and unspectacular. Slow, grinding work.'

That evening, 4 KSLI's carrier platoon, leading 159 Brigade, worked its way into the northern part of Vassy and reported that it 'was found deserted, very slightly damaged but completely looted, and booby trapped'.

The divisional history summed up the action during 14 and 15 August:

The local infantry actions necessitated by pockets of resistance and the searching of the woods around them made progress slow, and it was not until the second day that the immediate objectives of Vassy and the high ground to the north-east of it were secured. The bridge at Vassy was blown, but the stream there proved no obstacle to tracked vehicles, and the enemy on the farther bank withdrew upon the approach of our main forces.[4]

The Rifle Brigade's regimental history described the 'indicators' of German resistance that they looked out for during the advance to the Falaise Pocket:

As the advance gathered speed and the issue of the fighting began to be clear, the French civilians declared themselves openly and with the wildest enthusiasm, greeting the Riflemen with a mixture of eggs and kisses. No one who 'liberated' a French village will ever forget the Gallic abandon with which they were met. Apart from material rewards in the form of Calvados and cider, camembert and chickens, and an occasional glass of brandy, the inhabitants were a great assistance to us as we advanced. For one thing, they acted as effective mine detectors. If there were flags on the houses it was a sure sign that the enemy had gone. If one entered a village in silence, with no flags, no welcoming cheers, no children, only eyes watching silently from windows, stray dogs, a cart, perhaps overturned, then one could expect to meet the Germans round the corner.

16 August

The brigade's Op O No. 12 provided an analysis of the German situation in front of the day's planned advance,

The enemy is continuing his withdrawal and as yet shows little sign of losing control of his tps on our own front. 9 SS Pz Div have certainly been responsible for the rearguard actions on our left axis and are likely to continue to delay us until the R NOIREAU is reached. On our right axis the opposition was until yesterday provided by elements of 21 Pz Div. This fmn is now considered to have been pinched out by the narrowing of the front and we are likely to encounter tps from 326 Inf Div which has recently been reinforced by 988 GR of 276 Div. The enemy will probably attempt to stand temporarily on the high ground SE of the R NOIREAU, and will certainly delay us to the utmost with demolitions and scattered mines. It is unlikely that he will make any firm stand however until EAST of the FALAISE gap.

Brigadier Harvey gave his orders for the day at 0710 hours:

'Soft pedal' now off - objective high ground pt 218 V 8727 [Point 218 south of Conde sur Noireau]. On account of difficult and very close country it was only possible to move on one route ST GERMAIN 8433 - LE PONT 8529 - MONTILLI 8628. 2 F&F Yeo and 3 MONS leading.

Frustration at the slow progress and the number of vehicles on a single route led to the 8 RB and 23 H battlegroup being ordered forward from reserve 'to find a second route SE from VASSY 7932' but such were the narrow lanes and hedgerows they had to turn back.

The brigade war diary recorded that: 'No contact with enemy tps during the day but adv very delayed by mines, demolished brs particularly at the Mill 863297.' The sappers reported that the river near the blown bridge at Le Pont was too wide for the brigade's scissor bridge but a few hundred yards downstream in the mill area, where the river flowed in two courses, it could be bridged. Meanwhile, patrols of the F&F Yeo located a ford of the River Noireau a little

The advance to the River Noireau, 16 August 1944, and the diversion to the mill north of Le Port (insert).

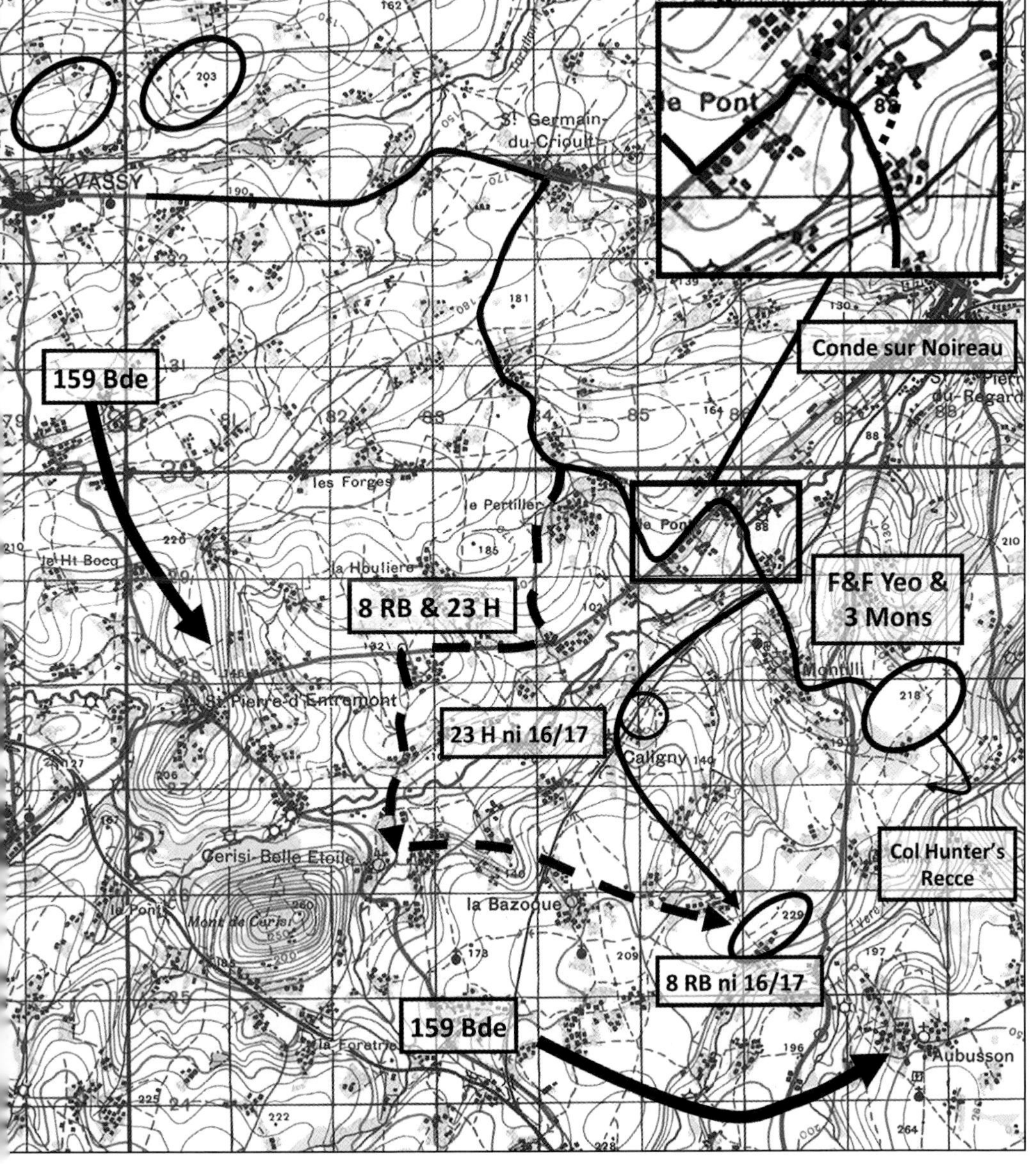

A Valentine scissor bridge. Each armoured brigade had one of these. The longer Small Box Girder Bridge mounted on AVREs was found in the 79th Armoured Division.

further upstream, but it was 'difficult'. However, it was just viable for tanks to cross and form a bridgehead. Later a troop of engineers improved the ford sufficiently for its use by wheeled vehicles. Once the scissor bridge was laid, vehicles had to cross water meadows before regaining the road. Consequently, this was the route subsequently taken by tracked vehicles, including those of 8 RB. Establishing these crossings and feeding the battlegroup across of course took time and would have been a very different matter if opposed. 'This was followed by a Bailey [bridge] constructed by 612 Fd Sqn and completed by 2200 hrs', using the abutments of the blown bridge at Le Pont.

On the other route, 159 Brigade was held as explained in 3 RTR's war diary at 1430 hours. 'Progress delayed by mines and felled trees blocking roads. Bridges at 822271 and 823271 on LEFT route cratered – considered impassable for several hours'. Consequently, with 29 Armoured Brigade having reached the Conde–Tinchebray road, a squadron of the 8 RB and 23 H battlegroup was sent west from reserve to Cerisy-Belle-Étoile, securing the crossing of the Noireau in the process. Both the F&F Yeo and 3 RTR battlegroups resumed the advance once the obstructions on their route had been cleared, but as dusk fell, 8 RB and 23 H

> were being urged to push on through the Third Tanks at their bridgehead over the Vêre River, a tributary of the Noireau. Colonel Hunter, of the RBs, and Major Blacker went on to investigate this for a suitable harbour area and found that it was indeed held – but by the enemy. In the end, we flopped down beside, in or under our tanks, on one of the dustiest roads in Normandy. We had put ten miles behind us.

By the evening of the 16th, 3 RTR and 4 KSLI had reached Aubusson but opposition in the village could not be cleared in the growing darkness and in the 29 Armoured Brigade area a mix of operational and logistic traffic blocking the roads brought 8 RB and 23 H's movement to the Vère to a halt. However, up ahead there was good news; an Inns of Court patrol had entered the town of Flers and reported it clear of enemy.

17 August

'Our first comparative 'swan' began … and continued for some hours.'

On the morning of the 17th, 159 Brigade Group passed through Flers and it was only at 1150 hours that the leading elements shook themselves free of an ecstatic population and pushed on along the main road towards Briouze. Meanwhile, at 0700 hours, 29 Armoured Brigade were on the move advancing on two parallel routes to the River Rouvre. In the war diary the adjutant of 8 RB recorded that:

> … the adv speeded up and apart from a few sharp engagements there was little to record … Each day the enemy withdrew voluntarily some 10 km, and each day we advanced at least that distance. Demolitions were frequent and mines were found, but no armd opposition was encountered …

In a trend that had been noted over the previous days, the number of prisoners of war noticeably increased, particularly Poles and other *Beutedeutsch* conscripted from Greater Germany. One prisoner obligingly confirmed what had already been deduced, that his erstwhile unit was withdrawing 15km (10 miles) every night, preparing defences on a river line to halt the enemy, before withdrawing again. Consequently, it was no surprise to find the Germans dug in on the line of the River Rouvre, where H Company and C Squadron swung into action. The river

> ran in a deep valley, the bridge was found to be blown. 14 platoon immediately crossed and occupied the high ground on the far side, followed closely by 16 platoon and then 15 platoon. Meanwhile the Squadron had

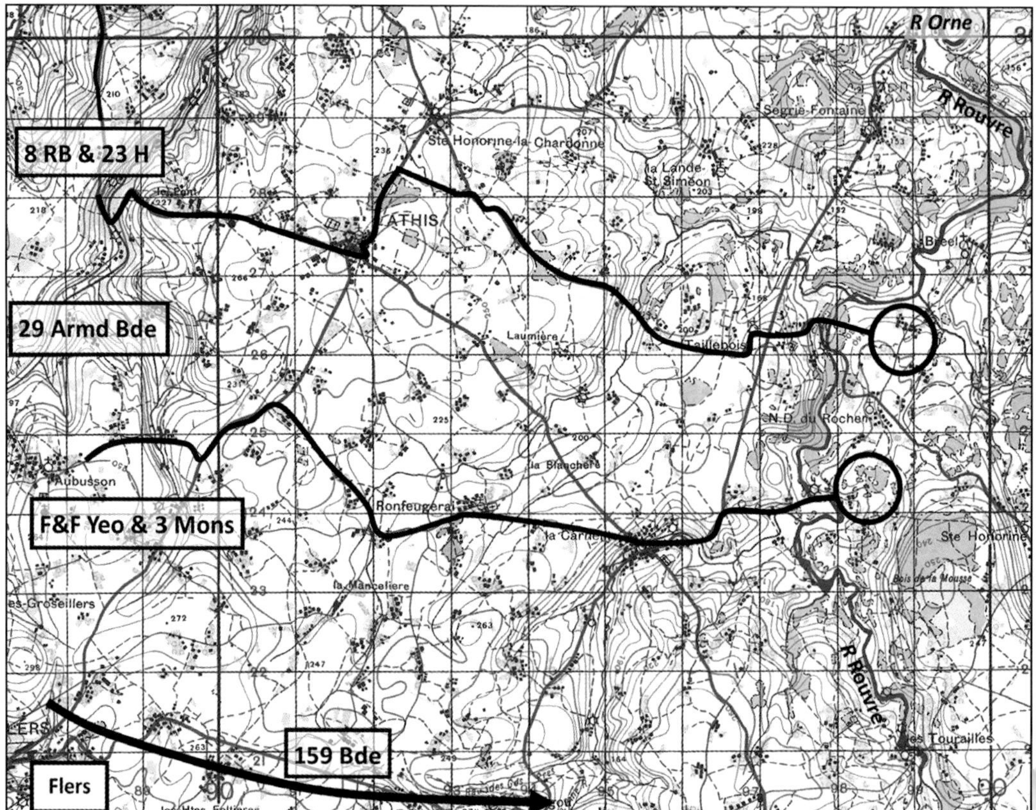

The advance to the River Rouvre on 17 August 1944.

found a place where they could cross their tanks, and shortly afterwards the entire group, less the scout platoon, who could not use the ford, were across the river and had consolidated the bridgehead.

With the use 'of brushwood and other forms of home-made bridging material', the sapper troop and riflemen made the river fordable for carriers, while they waited for trucks with bridging material. Major Bradford continued: 'The Sappers started building a new bridge at once and the site, as well as the bridgehead, was under 88 mm and mortar fire for the rest of the day. As soon as the bridge was completed G Company group passed over and resumed the advance.' Sergeant Hicks, with H Company, recalled that his company's:

14, 15 and 16 Platoons managed to wade across and secure the high ground on the other side of the valley and some tanks found a suitable crossing where the river was shallow enough for them. It was remarkable how quickly the Sappers appeared on the scene and I recall watching them struggling to erect sections of bridging materials in very hazardous conditions as the whole of our area was under mortar and shell fire throughout the day. I went over the river and up to the top of the high ground where I found a position overlooking the German-held area. I located one spot from where the rocket mortars [*nebelwerfers*] were being launched and was able to give it a bloody good plastering with our 3 inch mortars. Must have done some good as the amount of stuff we were receiving noticeably slackened off.

Having understood the Germans' withdrawal tactics, Royal Engineer officers and senior NCOs were forward with the leading motor company and under the protection of the infantry recced bridging sites and passed back information to the squadron, which again deployed the brigade's Valentine scissor bridge or dispatched trucks loaded with Bailey bridging as required. Keeping the sappers' trucks forward, although using valuable road space, helped maintain the battlegroups' momentum. With G Company group across the bridge, 8 RB recorded that as usual they came up against the German rearguard in Notre-Dame-du-Rocher that 'provided some inf opposition but by last light the position was cleared up'. The company's account provides more detail of the attack on the village

> by 10 and 12 Platoons, supported by A Squadron, 23rd Hussars. The enemy withdrew, but 10 Platoon were mortared on their objective and 9 Platoon had suffered some casualties from sniping, after successfully occupying some farm buildings which overlooked the village. Eventually 10 and 12 Platoons came out of the village and 11 Platoon, with the leading troop of tanks and section of carriers, continued the advance. They very soon ran into trouble [at Notre-Dame-du-Rocher] and, as always seemed to happen, at this juncture we had to call it off owing to failing light. Our night position was mortared and shelled continuously, and more casualties were caused.

The Rouvre was not a wide river, but its high, steep banks and wooded approaches made it a significant obstacle to armour.

18 August

We now began to get into the area where the RAF had been strafing enemy vehicles and from then till our next rest there was an average of one wrecked enemy veh every 100x [yards] of road.

The brigade's objective for the day was to drive in a more easterly direction to Putanges and the River Orne. With the enemy having withdrawn from Notre-Dame-du-Rocher, 'B Squadron and F Company went through [G Company] and swanned merrily on, meeting no opposition.' By 1145 hours they had advanced 12 miles and reached the high ground overlooking Putanges, on the Orne. However, as they approached the area, a loud explosion was reported and a few minutes later it was confirmed that the bridge at Putanges, the Germans' last intact crossing of the Orne, had been blown.

The leading company/squadron groups began to search for a crossing either side of Putanges. To the north, approaches to the river were covered by a strong enemy rearguard in the village of Launay. As 13 Carrier Platoon approached a crossroads outside the village, Sgt Kitson's carrier section was ambushed and all but the drivers were killed or wounded in this first contact of the day:

> Capt. May then took a second van van-guard [carrier section] round to the right and directed on LAUNAY, where they again encountered stiff opposition. The group was not strong enough to make any further progress and the rest of the day was spent in taking offensive action to stop the enemy infiltrating on to the Brigade C. L. That night the group, less one van van-guard, sat on the cross-roads just outside LAUNAY while the van van-guard was on the crossroads where first contact had been made. 15 Platoon also suffered fairly severely as a result of this action.

The citation for the Military Cross that Captain May, commanding half of H Company that day, received for this action adds more detail:

> On 18th August, 1944, Capt. May was in command of the vanguard, consisting of a section of carriers and a mortar platoon of H Company, 8th Rifle Brigade, supported by a troop of tanks of the 23rd Hussars. West of Launay enemy opposition was met and heavy fire from artillery was opened on the vanguard. There was also considerable small arms fire. Capt. May organized an attack on the village by the infantry, supported by the tanks. During the attack, Capt. May saw that the supporting fire from the tanks was ineffective, and he immediately jumped on to the troop commander's tank to secure better supporting fire.
>
> All this time the tanks were under machine-gun fire and Capt. May was completely exposed. Shortly after this, two riflemen were killed in a farmyard. Disregarding his own safety and in spite of enemy small-arms and mortar fire, Capt. May went forward by himself to see whether the men were wounded or killed.

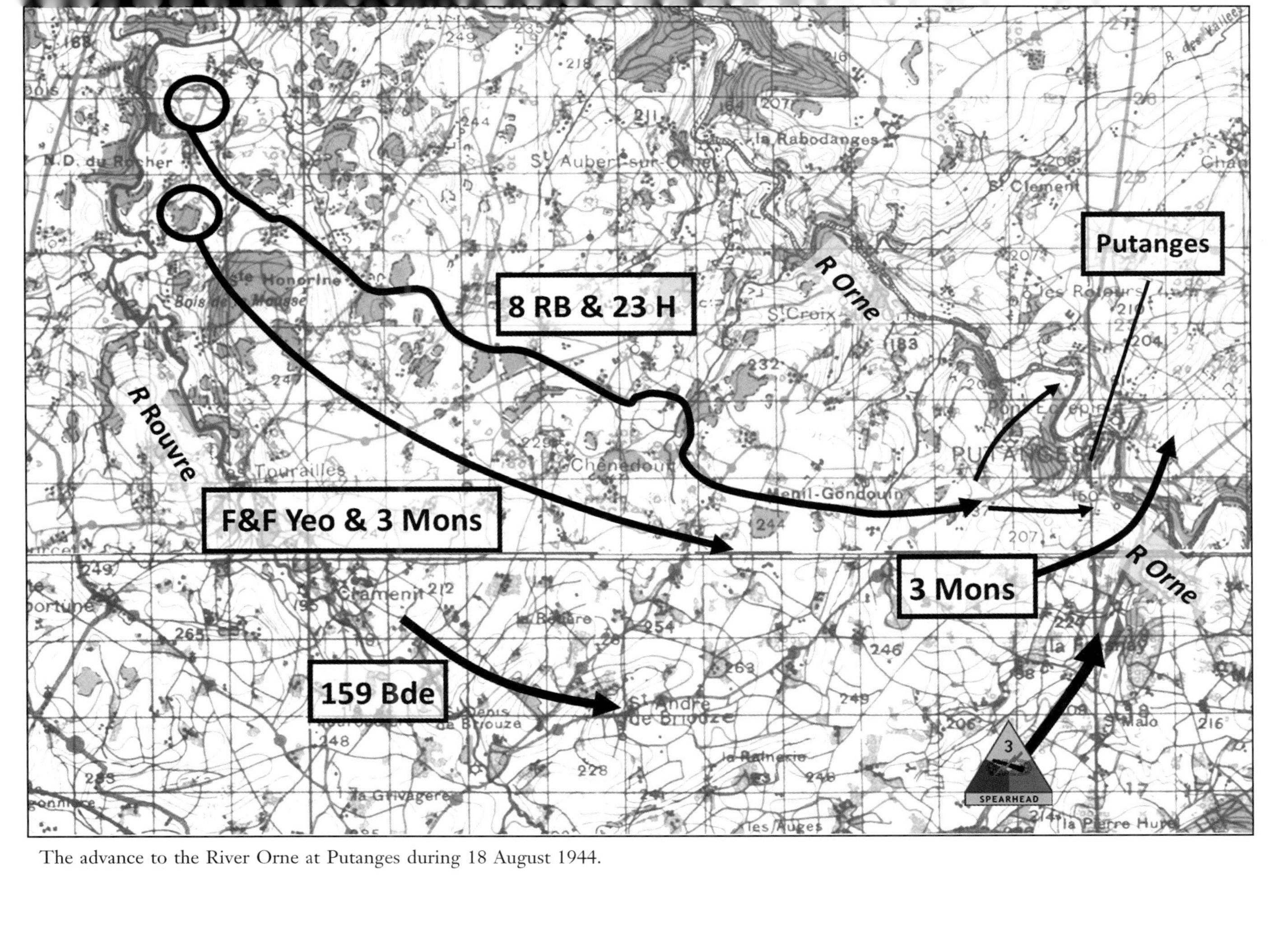

The advance to the River Orne at Putanges during 18 August 1944.

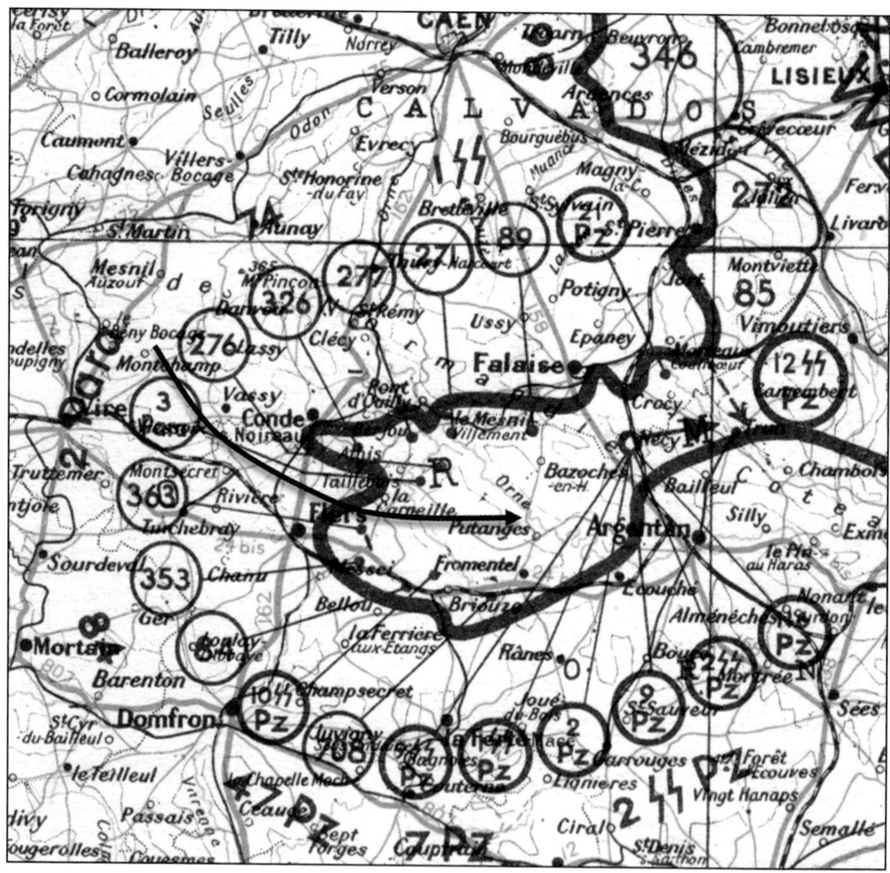

An OH map showing the remnants of the Seventh Army being compressed into the Falaise Pocket in increasing confusion. Photographic evidence shows that at least some of the time during this phase of the advance, 8 RB were facing the 3rd *Fallschirmjäger* and the 363rd divisions. Missing from the map is the 21st Panzer Division, elements of which were encountered by the 11th Armoured Division.[5]

The infantry under Capt. May's command continued to attack and harass the enemy with great effect until ordered to withdraw by his company commander.

Heavy casualties were suffered by the vanguard in this action and Capt. May's gallantry and complete disregard of danger were an inspiration to the men under his command.

Meanwhile, G Company was with Major Bell, who recorded that:

At last light a force under my command, consisting of 11 and 12 Platoons, a machine-gun platoon from E Company and a troop of self-propelled 17-pdrs, took up a position overlooking the proposed bridgehead, the object being to deal with any enemy who might try and counter-attack in the morning. The remainder of the Company remained where it was under

Kenneth Chabot. No counter-attack developed, however, but the machine-gunners found an excuse to open up and, as was always the case, once they had begun there was nothing in the world which would persuade them to stop.

Despite the German army reaching its denouement, many of its soldiers were fighting hard. As a measure of this, H Company alone suffered six dead and nineteen wounded during 16 and 17 August.

Elsewhere on the river line '… Sergt. Cooper made a reconnaissance of the River Orne with great skill and coolness and, in spite of the presence of two Panther tanks, returned with very useful information.' For this and other actions during the campaign he received the Military Medal. In the meantime, F Company and B Squadron on the southern flank had run not into the enemy but a column of tanks from 3rd US Armored Division on the same road to Putanges, which took all afternoon to sort out. The divisional historian wrote:

> By vigorous patrolling a limited crossing [of the Orne] was found to the south [of Putanges] over which a few tanks and some infantry were passed [H Company on another improvised brushwood crossing]. The Germans appeared to be in about battalion strength and opposed any approach to the main Putanges bridge by machine gun fire from the east bank.

As bridging could not begin until a bridgehead was established, during the afternoon 3 Mons were ordered forward to cross the Orne that night and establish a bridgehead on the high ground to the east of Putanges, which would enable the sappers to start work. There were concerns but risks had to be taken.

Royal Engineers launching a Bailey bridge across the River Orne.

'There was some doubt whether in view of the strength of the enemy, one battalion would be sufficient ... but it was appreciated that the enemy troops were not of sufficient calibre to withstand a determined attack, and in any case would probably withdraw once more after nightfall.' The operation began at 2300 hours with the rifle companies wading across the river, largely unopposed, aided by a string of toggle ropes. 'Light cas, inc one drowned.' By 0340 the Mons were across and digging in and a squadron of F&F Yeo managed to ford the river after dawn.

The Falaise Pocket – 19 August

The divisional historian summarised the overall situation as the brigade advanced to squeeze the southern flank of the Falaise Pocket:

Such delay as he [the Germans] could impose upon us, however, was vital to him for the extrication of his forces now facing the danger of encirclement east of Falaise. 50th Division were now brought in behind us; but in fact the opportunity to increase further the pace of the advance by leaving pockets of resistance to them did not present itself, for such pockets were invariably encountered astride all the few available routes, and had, therefore, to be liquidated before we ourselves could continue.

The advance was to continue on two centre lines. In the south 159 Brigade moved on the main road via Écouché towards Argentan, while 29 Armoured Brigade took a route 3–5 miles north, again in the direction of Argentan. At 0830 hours Brigadier Harvey held a CO conference. The aim for the day was to advance some 12 miles to the wooded ridge of the Fôret de Gouffern, where 23 H were to establish a firm base at Point 195, a dominating hill just east of the main Falaise–Argentan road. The rest of the brigades' objectives, advancing on two centre lines, were on the north-east side of the forest overlooking the pocket. This eastwards advance would cut across the front of XV US Corps that were advancing north and required some careful coordination.

The advance did not begin until 1200 hours but first to the west of the Orne, as the Germans had fought so hard the previous day, it was somewhat of a surprise for an H Company patrol to find that the Germans had abandoned Lunay. It was, however, no surprise when they went forward to the river with a troop of tanks to find that the bridge had been blown. The company squadron group was called back to the battlegroup, where they became the reserve.

The Bailey bridge at Putanges was recorded as being fully completed at 1500 hours, but the tanks and half-tracks of the battlegroup were able to cross an hour earlier when the bridge deck had been secured. F Company and B Squadron led the battlegroup on the road to Argentan in the pouring rain on a frontage of two platoons/troops and soon caught up with a German horsedrawn column, as described by Sergeant Cooper:

Leading the advance behind a fast retreating Hun Army, my section [of carriers] this day had been fairly quiet. Suddenly from seemingly nowhere,

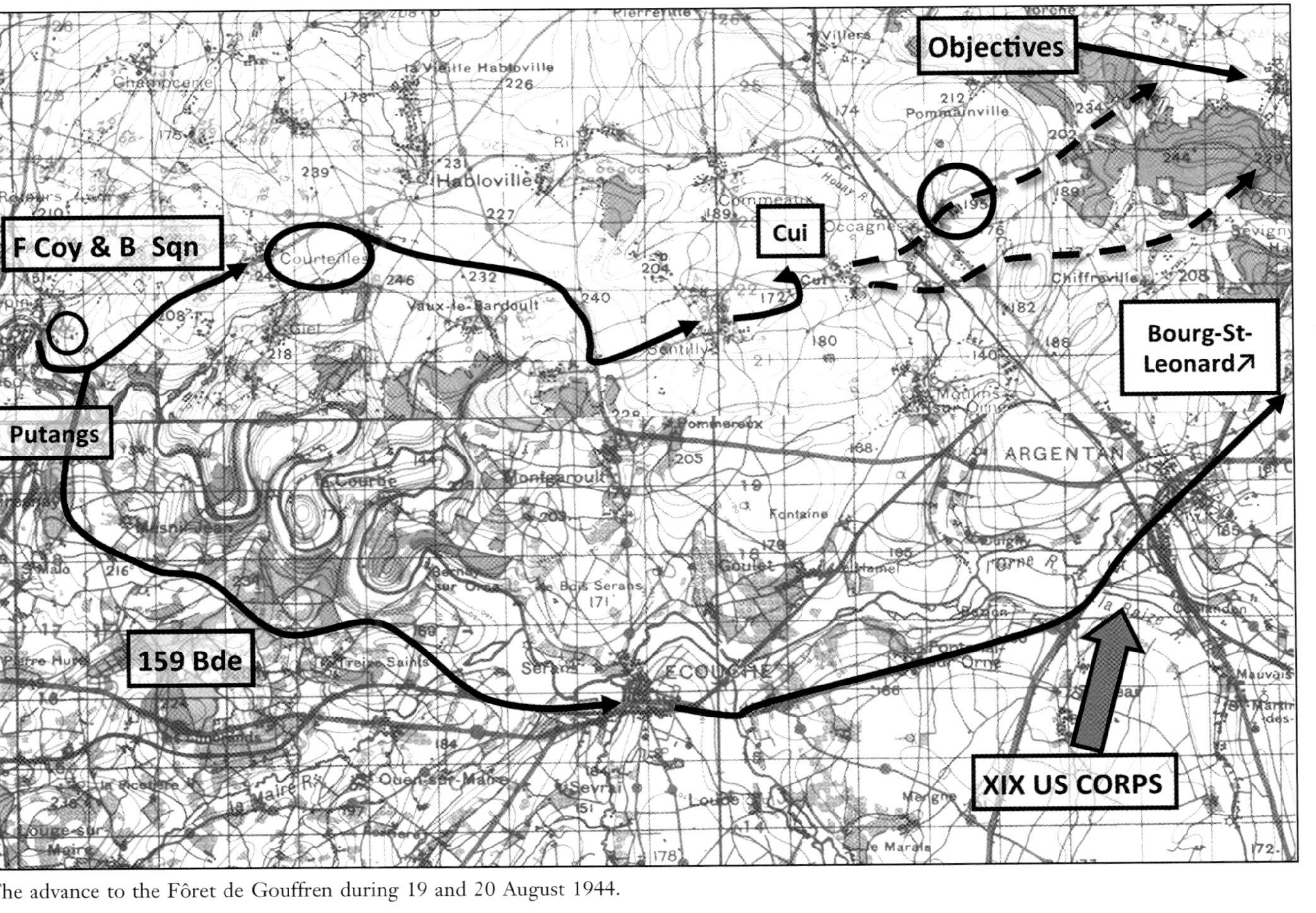

The advance to the Fôret de Gouffren during 19 and 20 August 1944.

actually by parachute from a Recce plane,[6] this message came 'Enemy horse-drawn vehicles approaching Centre-Line'. My carrier crept forward beside the leading vehicle to warn the Commander. As I began shouting at him, the first Jerry vehicle appeared about 40 yards in front of us. Quickly the leading carrier's gunner 'nipped' the horse, then all was a shambles. Knocking the next one off its feet, thus partially blocking the road, we went forward on foot, and round a bend beheld the amazing spectacle of about 30 wagons thundering down in line; the drivers whipped up their horses hoping to force a way round. This was a veritable Wild West Show. Chaos reigned while we stopped the charge, aided by an old Frenchman, who insisted on running around with a Cognac bottle and glasses. When the smoke cleared, much to our joy I counted no casualties to us, and I heard my Wireless Operator still giving a running commentary on the action.

When the advance resumed onto the first objective, the plateau of high ground around Courteilles:

Suddenly 8 Platoon of F Company working with Fourth Troop spotted two Panthers in a field. In the pouring rain, Major Wigan came up to place the seventeen-pounders. Commanders were out of their tanks guiding them into position. The RBs placed themselves along the road. But the Panthers showed no sign of life. After much discussion, they were approached and investigated and found to have been abandoned and put of action.

H Company and C Squadron had followed at the rear of the battlegroup, and as recorded by Major Bell:

Late that afternoon we passed through SENTILLY, where the leading tanks had a short engagement. Life was made a little more uncomfortable by some stray shots passing over us which were later discovered to have come from the 15th/19th Hussars [15/19 H] who were on our right rear.

They were lucky as following the incident with the abandoned Panthers 'a more active and mobile opponent appeared quite unexpectedly in the shape of the 15/19th Hussars!' This regiment had only arrived with the division on the night of 18–19 August to replace 2 N Yeo, who were being broken up to provide replacement tank crews to keep other Cromwell-equipped units up to strength. The N Yeo's historian noted that:

The 15th/19th Hussars arrived on transporters having left England less than a week before. A number of vehicles including technical stores lorries, fitters' half-tracks, scout-cars, etc, were handed over to them during the night. They were supplied with maps, codes, wireless frequencies, and as much information as possible. By 10 a.m. they were netted in and moving off in Divisional reserve without really having the faintest idea what it was all about – which was no fault of theirs.[7]

Up ahead, G Company and A Squadron, who had taken over the lead, had turned off the Argentan road, driven through Sentilly and an advance guard pushed on in growing darkness towards Cui. 'Here the leading tank was knocked out and thereafter a strange and uneasy situation developed.' Two of the Sherman's crew were wounded and taken prisoner.

With the number of Germans variously fighting on or bent on escape or surrender, the situation was tense and clearly fluid. Consequently, isolated near Cui, G Company, following an abortive tank-hunting patrol for the Panthers, were at around 2000 hours ordered 'to pull back from the village for the night and we retired to an area where we found a perfectly good, abandoned Panther. Here we "dossed down" for a few remaining hours of the night still left to us.'

The battlegroups had halted overnight still to the south-west of the Fôret de Gouffern, which was reported to be full of German troops, tanks and supply dumps.

British infantry advancing on foot. Even though 8 RB had half-tracks, they were not heavily armoured, so when necessary to protect their vehicle and the tanks in the bocage, they dismounted.

20 August

The fight to close the Falaise Pocket further east had been underway for days and 8 RB/23 H battlegroup were now in the main part of it, where the Germans were still doing their best to escape through the narrowing but rapidly closing neck. As far as the advance through the Fôret de Gouffern to the extant objectives on the other side was concerned, the brigade advance was slow, being constrained by boundaries to the south with the 3rd US Armored Division and control measures to the north. These were necessary to prevent the battlegroup from being shelled by friendly artillery or attacked by the Allied tactical air forces. In addition, 'The whole area was infested with Germans.'

That morning, G Company and A Squadron resumed the advance, as described by Major Bell:

> The position at Cui next morning was exactly the same as when we had left it the night before with the exception that this time we were not greeted by any hostile fire. The village contained a large hospital which was full of German wounded, and which seemed to have an enormous medical staff, many of whom I am quite sure acquired red-cross armbands a very short time before we arrived on the scene. There were large numbers of seriously wounded Germans and some German nurses, who when not on duty appeared to be practising their feminine charms most effectively and blatantly on the walking wounded cases.

The two wounded Sherman crewmen were found in the hospital and evacuated down the British medical chain. While the German prisoners were being secured, the other two company squadron groups sent patrols forward to the Falaise–Argentan road. However, approaching the forest, B Squadron came under fire and lost a tank to a pair of Panthers but, outnumbered in the resulting engagement, the panzers were in turn knocked out by B and C Squadrons' Shermans.

Now it was the turn of the infantry to advance to the forest, but the further 8 RB progressed into the Falaise Pocket, the grimmer the scenes of destruction became. Rifleman Jefferson of E Company, 17 Anti-Tank Platoon, described what he saw in his journal:

> We moved forward slowly. The whole scene became increasingly more indescribable. The whole area was strewn with dead Germans and horses and wrecked transport, guns and knocked-out tanks of every description. We had to pick our way through the wreckage. Sometimes there was no other way through the devastation but to drive over the corpses. The stench was overpowering. Prisoners were giving themselves up in their hundreds.[8]

Sergeant Hicks provides a more graphic description:

> The retreating Germans were mercilessly strafed by our air forces as they had no protection at all from the Luftwaffe. I vividly recall seeing the result of some of the damage and carnage our air force had wrought on a German

Scenes of destruction multiplied as 8 RB advanced through the Fôret de Gouffern.

column caught on a main road. It was not only littered with the debris of burning tanks and numerous other types of armoured vehicles and lorries but also scores of dead horses, which the Germans had been using to draw guns and ammunition limbers.

It reminded me of black-and-white photographs I had seen of similar situations in the First World War. You had to see it in the flesh, in Technicolour, with the road literally awash with blood and entrails, human corpses in grotesque positions and to cap it all a sickly stench over the scene like an invisible fog.

I don't know how I stopped myself from throwing up. I think we were all rather stunned by what we could see. The conversation was practically nil except for exclamations such as 'My God' and 'Bloody Hell' and such like.

The road was obviously impassable and we were glad to find an alternative route across fields and to get away from the ghastly mess.

At 1600 hours, G Company and A Squadron took over the lead in the forest:

> After crossing the Falaise–Argentan highway at Occagnes, we approached the Fôret de Gouffern with a certain amount of caution, as several clearly camouflaged objects could be seen in the edge of the wood. Under cover of smoke, and after we had brought down a few precautionary rounds of gunfire on the wood itself, a patrol went forward and fired a P.I.A.T. bomb at one of these objects. A direct hit was scored, but on clear examination the object turned out to be an abandoned German tank and the other objects we had seen in addition also turned out to be the same. Accordingly, a section of carriers moved forward, but the leading one went up on a mine and once again the advance was slowed down. The woods either side of the road obviously had to be cleared by the motor platoons on foot and this plan was set in motion. More traces of mines were found and blown down trees had been pulled across the road, but with the aid of our sappers these were cleared and the vehicles began moving forward again.

While the riflemen cleared through the forest, the tank crews of 23 H had more time to take in the scene around them amidst the shell-torn trees:

> Occasionally there was a burnt-out lorry or car in the ditch, and sometimes the blackened and twisted corpses of men were visible inside, while other crushed, torn bodies lay around. Again, and everywhere, there was the smell of burning and decay.

The battle group reached their objectives overlooking the village of Bailleul. Below was the valley, which had become the final killing area, scattered with 'the pulverised remnants of the German Army in Normandy'. It was not just the German dead and wounded that Rifleman Jefferson recalled coming across but individuals and groups of bewildered German soldiers:

> Some wore the unmistakable [runic] sign of the crack SS troops, still their arrogant selves. The prisoners were a nuisance to us. We simply bypassed them and sent them back in long columns. Bulldozer tanks tried to clear the roads. There was simply nowhere to push some of the German debris, it was piled so high. Literally thousands of German dead lay around.

During 20 August the divisional cage processed some 900 POWs, while 13 RHA's batteries forward with the battlegroups 'poured round after round into the pocket at targets identified by an Auster air OP aircraft'. One notable set of POWs exemplifies the extent of the dislocation of Seventh Army's command and control, when G Company captured their first German general:

> It was at this stage that 12 Platoon, moving forward on foot, captured [Lieutenant] General Kurt Badinsky, commanding 271 [*sic* 276th] Infantry

A Sherman (top left) avoids the carnage choking a road in the Falaise Pocket.

Division, and his entire staff. An old lady had rushed up to them and told them that there were some Germans in a certain farmhouse, whereupon, on closer investigation, a General was seen at the window and was immediately covered while the remainder of the Platoon surrounded the buildings. Staff officers, clerks, and orderlies came tumbling out and very soon the complete Divisional Headquarters personnel were assembled in the courtyard, being enthusiastically searched. The old General, who had no idea what was happening around him, and who had no division left to command, was particularly anxious to surrender to an officer and refused to believe Michael Anderson was one, and when I arrived on the scene, I had considerable difficulty in making him believe that I was not a non-com-missioned officer. He found it very hard to believe that dress of British officers in action was almost identical to that of their men, and when the Colonel arrived, looking equally dusty and shabby, he was never more exasperated. A request was granted to him to say goodbye to his staff and, after much heel-clicking and saluting, the first live German General we had ever seen was driven away to captivity, leaving behind him two very fine cars, which were very quickly snapped up.

These two cars were described as 'streamlined and crammed with arrogant and immaculate German staff officers', who were 'passed back to Brigade Head-quarters with little ceremony'. However: 'With great difficulty, we managed to extract the two cars from Corps Headquarters and the Colonel became the owner

A dismounted infantry section threading its way through the bocage.

of a resplendent Horch, while Colonel Hunter took over an equally magnificent Lincoln Zephyr.'

Harboured on the southern edge of the pocket, 23 H describe the night air being 'filled with planes and flares and the long rumblings and vivid flashes of explosions'. For G Company: 'The ceaseless pounding by our guns and aircraft of the remnants of the German Seventh Army in the pocket continued all the time, and for once that sort of sound was rhythm to our ears.' During 20 and 21 August the fight for the Allies to close the Falaise Pocket was over, with German counterattacks to facilitate the escape of the remnants of the Seventh Army having been defeated.

The Last Lap – 21–22 August

The soldiers of 8 RB were spared the very worst sights of the Falaise Pocket between St Lambert and Chambois, with Rifleman Jefferson not just referring to the improvement in the weather when he later wrote: 'We eventually got through and the sun shone brilliantly as we passed through the ruins of Argentan.'

With the Allied divisions having converged on the pocket, the 11th Armoured Division had been squeezed out of the line and were to transit east through the XV US Corps' sector, via Argentan, on XXX Corps' Route DIAMOND. This would place the division ready to lead Montgomery's pursuit to the Seine.[9] Major Bell noted that on the morning of 21 August:

> Morale was high and soared even higher when plans for the next day were announced. A peace-time march along good roads appeared the form and in this we were not to be disappointed. After passing through the battered town of Argentan we overtook the [Battalion's logistic] Echelon, which had raced to the fore, then on through Exmes to the neighbourhood of Gace …

Having passed through the 2nd French Armoured Division, the 'peacetime move' did not last the day, as having travelled 12 miles, at 1700 hours orders arrived from 29 Brigade for:

> 23 H and 8 RB to seize area pt 309 5941 moving by X rds 5145 - X rds - rd junc 528418 - AUTHIEUX 5639. This coln encountered light opposition area AUTHIEUX but reached area pt 305 5740 by last light where it harboured for the night.

Just to the east, 23 Hussars reported enemy holding Point 309, with the support of an 88mm gun. However, as recorded by G Company:

> Opposition was again conspicuous by its absence the following day, except for mines which had been laid at nearly every cross-roads we came to and blown-down trees across the road. James Ramsden and the carriers, however, found an excellent way round and we arrived much quicker than we had dared to hope on the broad highway to Laigle, up which we streamed in no uncertain manner, getting an enthusiastic welcome from the civilians all along the way.

The battlegroup harboured near Grace for the night, having travelled a total of 20 miles.

The following morning, 22 August, 29 Brigade led the advance of XXX Corps east towards the Seine. It was resumed at 0730 hours and, according to H Company:

> before long we had our first encounter with enemy mines. The enemy had blown a bridge in the centre of a village called Ste. Gauburge, and we tried to find a way round the obstacle. In the meantime, oil drums, sleepers etc., were collected in an attempt to fill in the hole, and the whole village turned out, complete with some very reasonable champagne, to watch us at work.
>
> Eventually a civilian showed Major Hagger the way round and we moved on. More mines delayed us soon though and the Engineers had to come up and remove them. Then after a short advance of 5 miles, against no opposition we stopped for the night near the village of RAI SUR RILE, just outside

L'AIGLE. The most noticeable feature of the day was how very much more pleased the people were to see us than the farmers of Calvados had been.

Having negotiated the usual obstacles, blown bridges and a certain amount of rubble, plus the odd crowd of enthusiastic French, and covered another 20 miles, at 1830 hours the objective for the day, L'Aigle, was reached. Here they met Jeep-borne reconnaissance units of a US corps entering the village from the south. After over three weeks' moving and fighting, with one brief day's rest, at 2200 hours it was confirmed that the division would halt there for some days.

By the time 8 RB halted at L'Aigle on 22 August, they were 25 miles beyond the Forêt de Gouffern and a third of the way to the Seine.

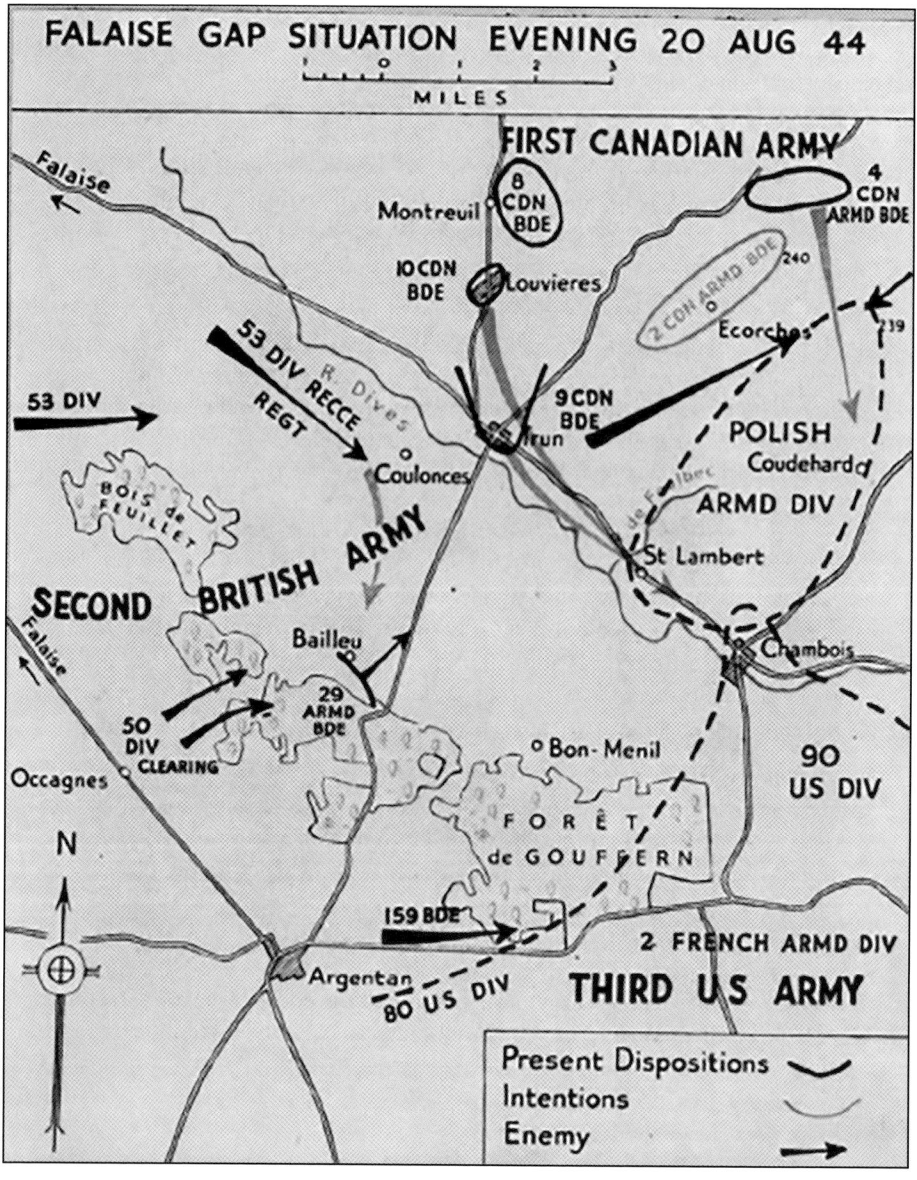

Chapter Twelve

The Great Swan

General Eisenhower confirmed that he would be taking over command of ground operations from General Montgomery on 1 September, but such had been the pace of events in the previous weeks, the Supreme Commander had few detailed plans beyond the adoption of the favoured US strategy of the day, the broad front. Despite the considerable distance and the logistical constraints of not having a significant port and in-loading across the beaches, Eisenhower's basic concept was for his six armies to line up on the Rhine before entering Germany.[1]

By the last week of August, it was clear to the Supreme Commander and General Montgomery that the collapse of the Seventh Army was more complete than could have been reasonably hoped for given its months of stubborn resistance. However, as early as 18 August, 21st Army Group had already been formulating a plan for the exploitation from in Normandy, as a part of Montgomery's subsequently vetoed drive deep into the Reich, the narrow front strategy. The first problem in either of the two strategies was crossing the important and substantial strategic barrier of the River Seine. Speed was clearly of the essence, giving as little time as possible for the Germans to mount a defence that would require a full-blown assault river crossing. It was hoped to 'bounce the Seine' or at the very least carry out an assault before the enemy's defences were fully constituted.

With the 11th Armoured Division halted at L'Aigle on 22 August, the remaining 60 miles of XXX Corps' advance to the Seine was carried out by the 50th Infantry Division, with the 43rd Wessex Division preparing to conduct an assault crossing and subsequent bridging of the river. Meanwhile, 8 RB and the rest of the division were reconstituting ready for a rapid advance from the Seine across northern France via Amiens, into Belgium and a 'very distant objective', Brussels or the port of Antwerp. Lieutenant General Horrocks noted that:

> Montgomery had made it clear to me that there must be no relaxation of pressure by day, nor, if necessary, by night. 'The Germans are very good soldiers,' he said, 'and will recover quickly if allowed to do so. All risks are justified – I intend to get a bridgehead over the Rhine before they have time to recover.'[2]

At Rai-sur-Rile, Sergeant Hicks recalled that H Company was:

> … spread out over a very large field. It was rumoured that we would be there for a couple of days only but in actual fact we spent five whole days there

Lieutenant General Sir Brian Horrocks took over XXX Corps (shoulder flash inset) after BLUECOAT in early August and 11th Armoured Division remained his under command for the advance from the Seine.

basking in glorious sunshine. It was most enjoyable and we took full advantage of this unexpected break from less pleasant things. Not only did the NAAFI van catch up with us, which was quite a novelty in itself, but an ENSA group appeared as if by magic and gave us some very welcome entertainment.[3]

During this pause, reinforcements arrived for 8 RB, along with replacement vehicles, which together as a high-priority unit more or less brought the battalion up to strength for the next phase of the campaign. It goes without saying that the REME detachment, regimental fitters and drivers had little time to relax in the sun, as they tackled a backlog of maintenance in preparation for the advance north-east from the Seine.

By 24 August the 43rd Wessex Division had reached the Seine and were concentrating for the assault crossing and the building of a Class 40 Bailey bridge code-named GOLIATH and DAVID, a Class 9 bridge.[4] The following evening

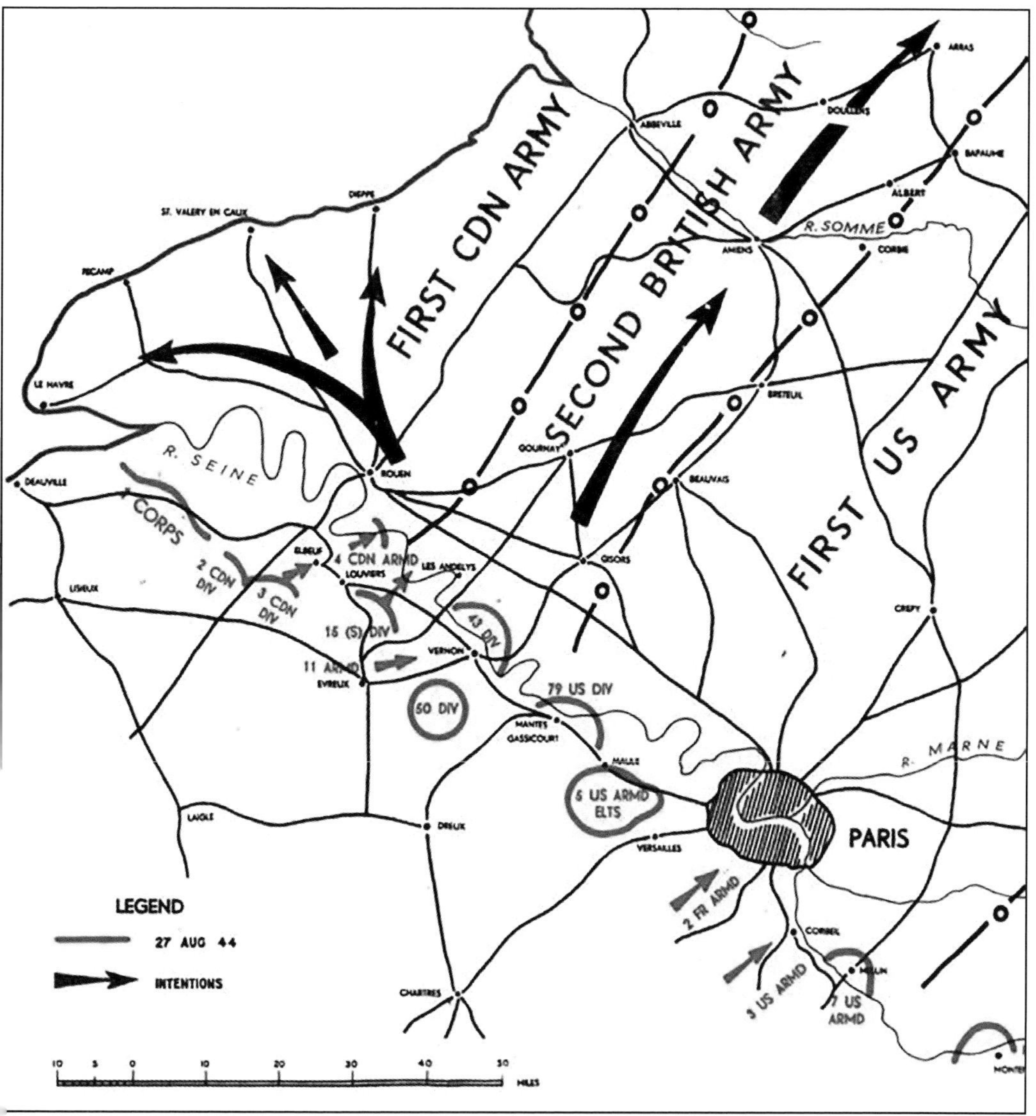

The Great Swan.

three of the Wessex Division's infantry battalions crossed the river in assault and storm boats to seize a bridgehead against relatively light resistance. With this success, HQ 29 Armoured Brigade issued a warning order at 2330 on 26 August, ordering the regimental groups to begin the 60-mile march to the Seine the following morning. However, it was estimated that once the bridgehead was secure enough to permit work on the bridges, it would take the sappers thirty-six hours for the Class 40 to be completed. Consequently, the brigade expected some time in an assembly area west of the Seine for servicing, resupply and rest, before beginning their armoured thrust across northern France.

For the type of fighting anticipated in the coming battles, the motor companies were again regrouped with their original armoured regiments and Colonel

Hunter's battalion headquarters resumed its role of moving at the rear with 119 Anti-Tank Battery. In this flexible use of his motor battalion, Major General Roberts was at this time leading the way. In other divisions, groupings tended to have become permanent.

With instructions for regrouping issued, Sergeant Hicks wrote:

> Orders to pack up and move on came all too soon but the rest and sleep, free from the noise of battle had greatly refreshed us all and we left L'Aigle on the morning of 28th August with quite a spring in our step. It was a long drive up to the head of the bridgehead at Vernon on the Seine which took us all day.

The division's leading vehicles were on the move at 0545 hours on the morning of 28 August, on extremely dusty roads during what was to be the last day of settled summer weather. The battlegroups arrived in their concentration areas west of the Seine between 1245 and 1630 hours, having suffered a very modest number of breakdowns for the length of the march. The brigade's expectation of at least a night in the assembly area was ruled out as GOLIATH had been finished ahead of schedule and despite the resistance of two fresh battalions of the German 49th Division, the Wessex battalions had been able to establish a sizeable bridgehead. Consequently, at 1600 hours on the 28th a warning order was received for another move that very night, this time across the river. While vehicles were hurriedly serviced, refuelled and checked over, the brigade's battle procedure was under way, with the commanding officers attending Brigadier Harvey's orders at 1700 hours. A summary of the orders was recorded in the brigade's war diary:

> Verbal Orders – Bde gp will cross SEINE tonight to conc area TILLY 4778. Prepared to adv an two routes to AIMIENS N. 1058 first light 29 Aug. 3 R Tks to move at 1830 hrs with under comd 'G' coy 8 RB 'H' bty 13 RHA, and tp 612 Fd Sqn area 4878: 2 FF Yeo with under comd 'F' coy 8 RB 'I' bty 13 RHA and tp 612 Fd Sqn area 4778. 23 H with, under comd 'H' coy 8 RB, 'G' bty 13 RHA area 4777. Main Bde with 612 Fd Sqn, less two tps, and 18 Lt Fd Amb area 4777. 8 RB and 119/75 A tk bty area 4677.[5]

The armoured cars of the Inns of Court Yeomanry were to advance ahead of 29 Armoured Brigade covering the whole of the divisional frontage, while the Cromwells of the 15th/19th H were to cover the left flank. This was a temporary reversion from an armoured back to a recce regiment's role.

With advance parties having gone ahead across the river into the bridgehead to lay out harbour areas around the village of Tilly, at 1830 hours 3 RTR and G Company leading the brigade headed for the bridges at Vernon. As dusk fell, the heavier Shermans and Sextons queued to cross GOLIATH, while the half-tracks, carriers and the regimental groups' lighter vehicles followed the marked route to DAVID. Sergeant Hicks with H Company crossed the river in pouring rain and full darkness at 2100 hours. He recorded: 'The Royal Engineers had done a magnificent job in building a bridge across the River Seine, which was supported on pontoons, and it was quite an experience driving over it at about 10mph and feeling the bridge undulate in the water which felt a little eerie.' As the columns of vehicles climbed up from the river to the high ground, evidence of the Wessex Division's recent battle was to be seen, in the form of a still smouldering panzer with several bodies lying on it.

The battlegroups arrived in Tilly between 0022 and 0200 hours with head-quarters 8 RB at the rear of the column being the last to arrive. While tank and vehicle crews completed their checks, the riflemen had to dig in before they could get any rest. Sergeant Hicks continued his recollections. 'There was very little activity from the Germans, and it was obvious they were in some disarray and retreating all the time, for which we were duly thankful, so we spent a quiet and peaceful night in the bottom of our slit trenches.'

DAVID bridge at Vernon built by the sappers of the 43rd Wessex Division, opened late on the 27th for that division's support weapons and essential vehicles to cross first.

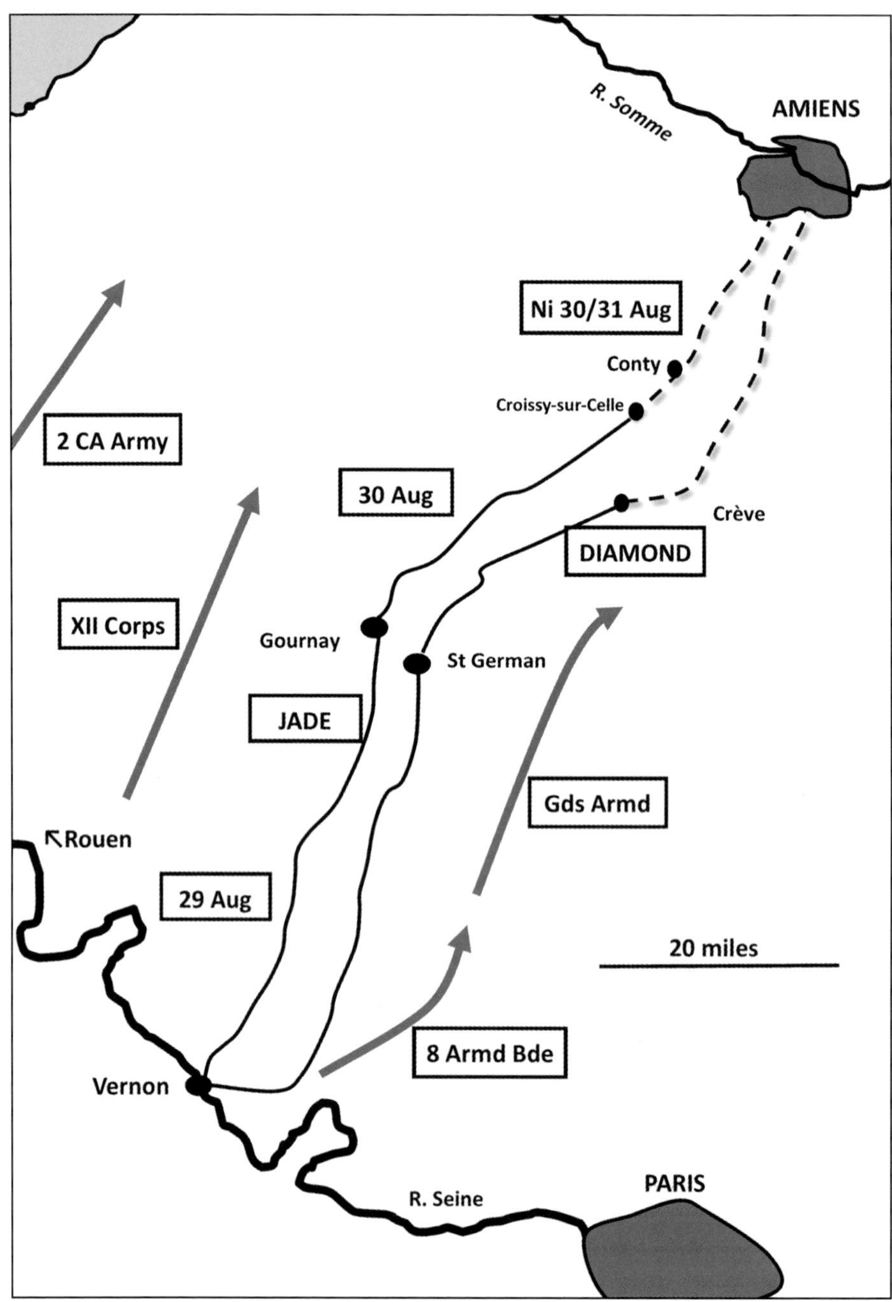

The plan for 21st Army Group's advance across northern France, spearheaded by two armoured divisions. Routes JADE and DIAMOND were allocated to 29 Armoured Brigade.

The First Day – 29 August

XXX Corps' plan for the next few days was to advance the 80 miles to Amiens with 29 Armoured Brigade to the left and 8 Armoured Brigade on the right. The latter would be leading on this route until such time as the Guards Armoured Division had crossed the river, caught up and taken over the lead on the XXX Corps' right. The 50th Division was to follow and be prepared to take on any serious or extensive defences that the enemy may have cobbled together and been bypassed by the armour.

Within the 11th Armoured Division, 29 Armoured Brigade would be advancing on two routes: 3 RTR and G Company right, F&F Yeo left and F Company, with 23 H and H Company in reserve. The instruction from Brigadier Harvey was that as much of the move as possible was to be across country. Still west of the river, 159 Brigade spent the morning snarled in the mass of traffic of a slimmed-down XXX Corps and only crossed the bridges during the afternoon of 29 August. Meanwhile, at 0700 hours, in heavy rain, the division's half-tracks and tanks were on their way to objectives a hitherto unbelievable 25 miles deep in enemy territory.

An entry in 29 Armoured Brigade's war diary recorded that the advance

> throughout the day continued against light opposition. It became apparent during the day that the enemy had little idea where we were, as the WEST - EAST rds from ROUEN [were] still in use by him until physically cut by our coln. Leading tps had good shooting at staff cars, lorries and horse drawn tpt on these rds.[6]

By the afternoon both of the leading regimental groups had encountered enemy resistance at Étrépagny. As recalled by F Company:

> Slight opposition was met on the first day of the re-continued advance at a place called Etrepagny, where a full Company attack was put in with some very welcome support from our 3″ Mortar detachments and also the tanks; this was too much for the enemy, who were trying to withdraw by the time we reached our objective.

What had happened is that as B Squadron approached Étrépagny they came under fire from a factory to the south-west of the town, losing a tank. With enemy infantry supporting several assault guns identified, F Company were brought up, while C Squadron were to attempt to bypass the village a good mile to the west. However, as F Company attacked the factory area, with the support of A Squadron, G Company entered Étrépagny from the east, presumably with the Germans having withdrawn, as they

> received a most enthusiastic welcome from the civilians who, however, suffered a bitter disappointment when, shortly after we had left, the Germans came back, and all rejoicing had to subside at once. However, it was not long before the situation was restored again.

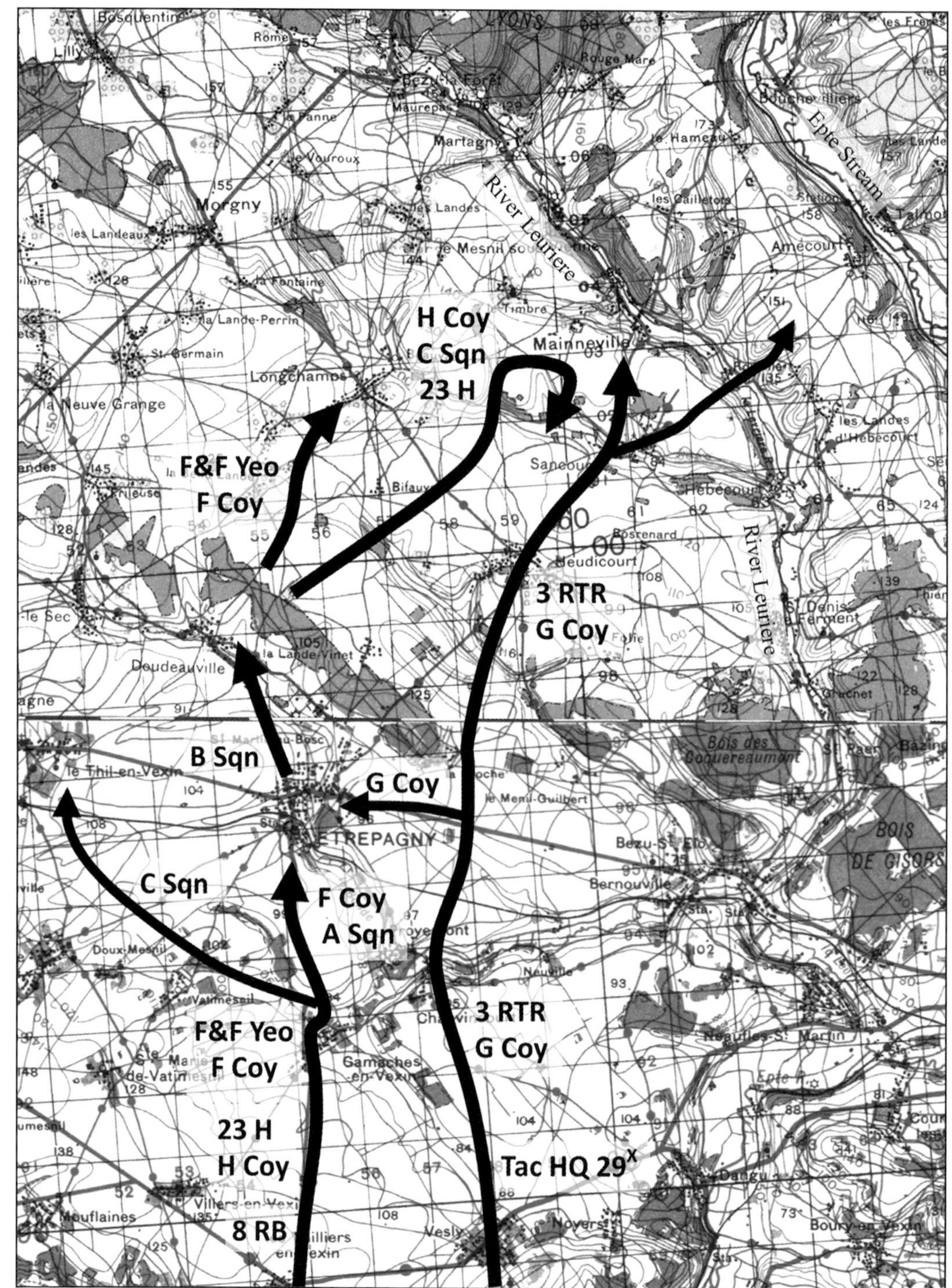

The advance of 29 Armoured Brigade on the afternoon of 29 August 1944.

The F&F Yeo's B Squadron pushed on through Étrépagny but by now the Germans were organised enough to mount further delaying operations and as the Shermans reached Doudeauville 'they again came into contact with anti-tank guns and had two tanks knocked out from the wooded area just north of the village. C Squadron endeavoured to get around the flanks. Two anti-tank guns were knocked out in the process and the advance continued to Longchamps.' Here the tanks encountered enemy infantry and in F Company's final act of the day they were 'mopped up'.

Meanwhile, 3 RTR and G Company were also pressing on but when approaching Mainneville with a view to crossing the River Lévrière, they came under fire and Colonel Silvertop ordered his squadrons to bypass the opposition via a bridge a mile to the south-east at Rouville. Here they also came under fire at 1830 hours. The river line was clearly a delaying position that was dealt with by tank fire, as half an hour later the war diary records G Company and some tanks were advancing onto the high ground of Point 151. Here they halted overlooking the valley of the Epte stream and the villages of Amécourt and Talmontiers, which were held by the enemy. Behind them, 23 H and H Company, which had been in reserve spending most of the day rounding up prisoners, were called forward to deal with the enemy in Mainneville. The company's historian wrote:

> After a lot of indecision as to whether we were going to attack or not, it was eventually decided that one platoon (Mr. Coryton's) would go forward and see what was actually in the village.
>
> He hadn't gone far down into it, though, before he heard Germans talking and grenades were thrown at the leading section. … it was decided, as we were too weak to deal with the opposition by night, to wait till morning.

That evening, Lieutenant Chabot, the company's carrier platoon commander, was ordered to take a motor platoon down into the valley to see if the Germans had withdrawn and check potential routes:

> I set forth with my party on a compass bearing, travelling on the leading carrier, but had some doubts as to where we should eventually arrive owing to the large amount of metal around being liable to render the compass inaccurate. However, I had more or less memorised the route before we started and knew that we had to leave a farm on our left. I was therefore very relieved when the farm loomed up in the darkness, but beyond it also appeared a number of objects which looked like vehicles. Our fears were set at rest, however, when on closer investigation these turned out to be bushes. We then went forward again and soon reached the point where we had decided to get out of our vehicles. My orders were to find out if there were any enemy in the village [Amécourt] and also to ascertain if the road through the village was suitable for tanks. On the map there was a track marked running straight down to the bridge, and I also had to find out if this was negotiable for vehicles.[7]

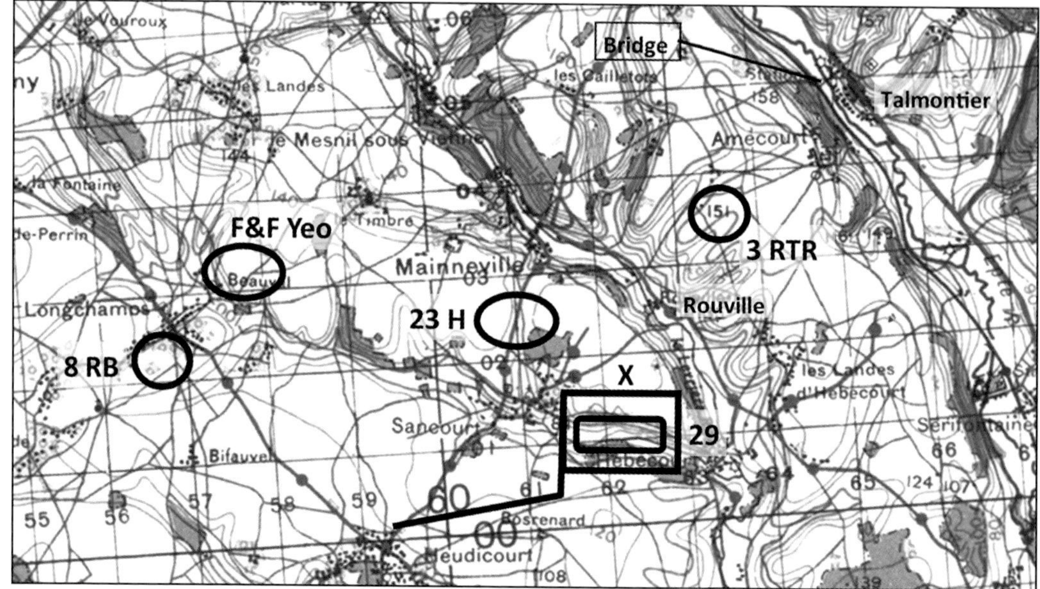

A section of the map of the Mainneville–Talmontiers area showing locations on the night of 29–30 August, GSGS 4249 Sheet 9E Neufchatel–Rouen.

I accordingly sent off a party from the carriers to go through the village and took with me two sections of 12 Platoon with a view to 'recceing' the track, and afterwards working up towards the village. The remaining section I left to look after the vehicles. I can say here and now that we never found the track that night, although we spent some four hours looking for it … In the darkness it was quite impossible to move quietly and I decided it was definitely 'not on' to approach the bridge through the woods. I therefore told Goldsmith to report this back on his 38 wireless set. He set about this task, speaking at the top of his extremely penetrating voice, thereby multiplying greatly the number of grey hairs in my head. He failed to get through … Most of the men with me were freshly out from England, and I can only think that they had had very little, if any, night training, as moving along I had the sensation of having a herd of elephants following me. Looking back on it now, I am appalled at the risks we took that night.

Brigadier Harvey was pleased and included the following in his Op O for the following day: 'The bde comd congratulates all ranks of the bde on their skill and determination in achieving a 25-mile advance on 29 Aug in the face of enemy and climatic opposition.'[8] The division in their first day of the Great Swan across northern France had become accustomed to brushing aside or bypassing the enemy, but more would be expected of them over the coming days.

The Second Day – 30 August

During 29 August, the Guards Armoured Division (XXX Corps) had closed up to the front to the right in order to take over the advance from 8 Armoured Brigade. The next major objective for the 11th Armoured Division and XXX Corps was the River Somme. Speed remained essential if they were to get there before the

Germans organised a defence, but Amiens and its bridges was over 40 miles away as the crow flies, which was expected to take two days.

At first light both 23 H and F&F Yeo dispatched recce patrols to find crossings of the Epte stream and at 0930 hours, the advance was resumed in heavy rain, with the first objective being Amécourt, which had been abandoned but it transpired that the village of Talmontiers, north of the Epte, was still held. It was far more strongly defended than assumed, as was soon found out when G Company's 01 command half-track was knocked out and Major McCrae wounded and evacuated for a second time. With the company and tanks having withdrawn, they brought the villages under artillery fire and 23 H's regimental group bypassed the enemy. B Squadron remained behind covering the village until 159 Brigade to dealt with Talmontiers.

Lieutenant Delaforce, an officer in 13 RHA, described the gunner's role in supporting the advance from the Seine bridgehead:

> The two artillery regiments, 13 RHA and Ayrshire Yeomanry, deployed into action several times a day. But by contrast with Normandy, the necessary fire support was usually quick troop or occasionally battery targets, and then it was 'up sticks and on'.[9]

The American-built M3 Stuart was known as the Honey in British service. Most had their turrets and 37mm guns removed earlier in the campaign but retained a machine gun.

During the advance from the Seine, the divisional artillery was reinforced by a single battery of guns from 64 Medium Regiment, detached from 5 AGRA.

Passing through G Company, H Company's historian noted very few incidents on their route to Crèvecoeur. They

> moved into a small village on a cross-roads. As we did so, a German lorry also moved in, and Captain Weiner's 'Honey' shot it up. We also captured a Ford V-8, which was unfortunately crashed at a later date. We didn't stay there long though, and moved on to MARSEILLE-EN-BEAVAISIS.

At Marseille-en-Beauvaisis, G Company and 3 RTR resumed the lead and by late afternoon on 30 August the regimental groups were well on their way to the day's objectives against little opposition.

'It's moonlight tonight'

At about 1600 hours General Roberts, forward in his tactical headquarters, received the warning order, 'It's moonlight tonight' from the corps commander and that General Horrocks was in his tank and on the way to see him. The advance from the Seine was one of the very few times in the North-West European Campaign that a British corps commander conducted the battle mounted in a tank. Horrocks recalled:

> As a result of my experience when in command of the 9th Armoured Division during exercises, I realized that the only way to command a formation consisting mainly of tracked vehicles was to be in one myself. In this particular case it was obvious that my Corps HQ, with its many soft-skinned vehicles, could not possibly advance until … the small centres of opposition had been mopped up. Unless I could move freely among the leading formations, I would soon be completely out of the picture and quite unable to 'smell the battlefield'. In a fluid operation like this it is important for the Corps Commander to be able to visit and talk to Commanders personally. My visit to Roberts's HQ is just one example of this … it was important that I should meet 'Pip' Roberts, the Divisional Commander, personally, and discuss the situation with him, so as to make sure that his Division was capable of carrying out this extremely hazardous operation.

The phrase in the warning order, 'It's moonlight tonight', from one desert veteran commander to another, was readily understood by General Roberts as being orders for a night advance on Amiens. While the operations staff plotted the route, the division's logisticians faced a serious challenge as they were working off lines of communication that stretched 50 miles to the west of the Seine. Even with priority on the two bridges, they had to fight to get 29 Armoured Brigade's A1 Echelons, reinforced by RASC petrol bowsers, forward through traffic on the two routes to deliver the vital fuel. The divisional provost company performed miracles, despite numerous arguments with angry commanders who had been ordered off the road to let the trucks through. Even so, not all 8 RB's companies received a full resupply.

With the arrival of General Horrocks at Roberts' Tac HQ, radio orders were passed down to 8 RB and the brigade's other units confirming that the current advance was to be continued until last light, with the final 20 miles to Amiens (nicknamed TABRIZ) being covered during the night of 30–31 August. This was to be followed by a dash to seize the Somme bridges in the city before the Germans could blow them. As recalled by Sergeant Hicks, after a day on the road Major Bell broke the news to his company of the change of plan:

> It was late afternoon, and we hoped we would be able to settle down there the night. Food was prepared and eaten and we generally tidied ourselves up after what had been a tiring day. I went to the usual 'O' Group for orders for the next day only to learn that there would be no rest for us that night. Gen. Horrocks had decided to push on to Amiens some 30 miles away to seize the bridge over the River Somme. We loaded up with all the spare petrol we could get hold of …

The first phase of the new plan was to continue the advance for another 25 miles to within striking distance of Amiens. At 2130 hours, with 29 Armoured Brigade halted to refuel, once again General Roberts changed the grouping of his armoured brigade to reflect the challenge of fighting in an urban environment where infantry is at a premium. For 8 RB this meant that the motor companies, less G Company, left the armoured regiments to come back under command of Colonel Hunter in order to create a body of mobile infantry that could be

'Orders'. Officers of 8 RB are wearing brown berets and tank officers black. Major Bell is on the left.

committed to battle as a battalion or as companies as dictated by the situation. General Horrocks commented on the changed regrouping within the division for the final phase of the advance to Amiens:

> One of the interesting things about this advance was that the two armoured divisions [11th and Guards] operated quite differently. The 11th Armoured was more flexible of the two. 'Pip' Roberts, one of the most experienced tank commanders in the Army, would switch his troops rapidly, attacking with tanks and infantry, or infantry supported by tanks according to the situation. The guards on the other hand always worked in regimental groups [e.g. Irish Guards tanks and infantry] ... This system was not calculated to get the full value from the mobility of an armoured division but ensured the closest possible cooperation between the two arms ...

The divisional plan given out as radio orders was summarised by General Roberts:

> Instead of the armoured brigade leading on both routes with the infantry brigade following, brigades would be organised as follows: on the right route, which was the main road from Crevecour to Amiens, would be 29th Armoured Brigade in the following order: 3rd Royal Tanks [and G Company] 8th Rifle Brigade, 29th Armoured Brigade HQ, 13th RHA, 23rd Hussars, 4 KSLI, Main Division HQ, Medium Battery RA. The left-hand route was a winding country road through Conty and Taisnil to Amiens and on this route would be 2nd Fife and Forfar Yeomanry, [8th Rifle Brigade (–)], 3rd Monmouths, HQ 159th Infantry Brigade, Ayrshire Yeomanry RA, 15/19th Hussars, 1st Herefords. The leading troops on both routes would start moving at once, and adjustments in organisation would take place as the advance continued.[10]

This plan ensured that there would be sufficient infantry forward should the Germans put up a stout defence of the city. The plan was duly disseminated to the companies and squadrons. Major Bell of G Company recalled:

> Company HQ commandeered a house for a meal, and the eggs were still sizzling in the frying pan when Colonel Silvertop [3 RTR], whose personal leadership inspired us all, received the order couched in the now famous phrase, 'it's moonlight tonight'. The parlour where we were cooking was the most suitable place to give out orders, and so, hastily eating what we could, the room was cleared for the order group, which assembled in this typical little French cottage kitchen, dimly lit by the light of lanterns. Only one question was asked, 'What do we do when we get to Amiens?' and this was later answered by the Brigadier by the words, 'Knock at the door and see if they will let you in.' A troop of tanks of the 3rd Royal Tank Regiment, was to lead, followed by G Company complete, with the remainder of the 3rd Royal Tank Regiment tailing on behind. The distance was approximately twenty miles and only one village lay on our route.

This village, Fontaine, was recced by the signals officer of the RTR and Lieutenant Ramsden of G Company, who went forward in a scout car and established that it was not held by the enemy.

Meanwhile, despite the fact that not all vehicles had received adequate fuel, the advance to Amiens was resumed at 2300 hours but not in moonlight as had been hoped but in pouring rain and 'impenetrable darkness'. On the right route, 3 RTR and G Company led, while the F&F Yeo followed by 8 RB were at the head of the left column. This route was the D8 road that led north along the valley of the River Selle, which was winding and, consequently, the slower of the two routes.

Night advances to contact with the enemy by armoured formations, without night fighting aids, were very rare during the Second World War for the reasons explained by Major Kershaw, Brigadier Harvey's brigade major:

> In an armoured vehicle one is essentially blind at night, especially where there is dust or rain and of course that primary sense for night fighting, hearing, is drowned out by the roar of engines. And pity the platoon commander at the front of the column who has to navigate by the light of a shaded torch, while behind him others attempt to follow the oft obscured convoy lights.

All of these problems are more than apparent in Major Bell's account of the advance to Amiens on the night of 30–31 August:

> It was altogether a strange and at times, frightening sensation to be ploughing on through enemy territory in the dark and having not the faintest idea of what might be in our path. Everyone was feeling very tired and I had to get the sergeant-major to ride up and down the column in a scout car to make sure the drivers of carriers in particular did not fall asleep at the wheel and cause a hold up of the column.

Trooper Buchanan of the F&F Yeo recalled the result of one navigational error up ahead of 8 RB as they made their way, without lights, 'more or less nose to tail'. 'One laugh was the lead tank, the boy [presumably a troop commander!] must have been bamboozled, instead of turning right, he went straight into a field! Of course, the whole British Army followed him.' [11] Another map reading error in C Squadron was recalled by Major Hutchinson:

> Whoever was leading the columns had got lost – we were driving backwards and forwards through the place. The odd thing was, it was still occupied by the Germans, but they had their heads down while we were messing around. Next day there had to be quite a considerable attack put in before they were cleared out.[12]

During the night there were numerous instances across the brigade area of German transport, in their efforts to get east to the Somme under cover of darkness, crossing routes or joining the slow-moving columns of 11th Armoured Division:

> In general, enemy opposition on the way was either brewed up or run over. German vehicles joined the column from side roads and sometimes travelled several miles with us before we realised their presence. Coloured lights were always being reported ahead, and there is no doubt that many of us that night were quite convinced that we had seen various objects which in point of fact never existed. At one time a large dark object appeared in front of us and the leading tank fired a round at it. The most gigantic explosion occurred, and the object, which turned out to be an armoured ammunition carrier, dissolved into small fragments. At the sound of the explosion a German was seen to spring up near my half-track and run faster than I have ever seen anyone run before … and to this day we like to think that that German is still running and unable to stop himself. At any rate, we were able to relieve our somewhat frayed nerves at this time by a good laugh at this unfortunate Hun's expense.

Not all encounters resulted in bloodshed, as recounted by Sergeant Hicks:

> We found German vehicles joining our column from side roads thinking we were a German convoy. During the many stops that occurred, the call of nature had to be answered and it was on one of these occasions that Charlie Large (the driver of company HQ's 02 half-track) [company 2iC's half-track] nipped round the back of his half-track for a piss and stood next to another bloke who was also relieving himself and it was only when the other bloke spoke to him in German that he realised who he was in the darkness. Apparently they were both so scared that they fled in opposite directions, with Charlie leaping back into his driving seat. No shots were fired because nobody could see anybody to shoot at. It was all a bit of a farce.

One of the division's Jeeps, with resistance fighters and Prisoners of War.

Chapter Thirteen

The Amiens Bridges

By 0400 hours on 31 August, 3 RTR and G Company were 3 miles short of the city, and on the left the F&F Yeo with the rest of 8 RB behind them were about 5 miles from Amiens. General Roberts recalled that:

> At this point, I gave order to 29th Armoured Brigade that 3rd Royal Tanks with the company of 8th Rifle Brigade to push on with all speed to Amiens and aim specifically for the main bridge over the River Somme; 23rd Hussars were to occupy the high ground south of the city astride the main road, and the 2nd Fife and Forfar Yeomanry on the left, the other main feature west of the little River Celle. The reverse slope positions south of 23rd Hussars and south-east of 2nd Fife and Forfar Yeomanry were to be held by 4 KSLI and 3rd Monmouths respectively. Also the CRA was to arrange for the medium battery RA to go into action positions, tying up with the divisional artillery. 159th Brigade were informed that their third battalion, 1st Herefords, would probably be required to help in the city.

At dawn, 0630 hours, the brigade's war diary recorded that:

> The adv on AMIENS continued throughout the night – no organised opposition was encountered but some horse-drawn tpt crossing the route was shot up [by 3 RTR]. Speed and density for advance – five MIH, density nose to tail – tail lights only were permitted.
>
> By first light units were disposed:
>
> | | |
> |---|---|
> | 3 R Tks | in outskirts of town waiting to enter |
> | 23 H | Area of DURY M 0853 |
> | 2 FF Yeo | still moving on left route |
> | 8 RB | immediately behind 2 FF Yeo |
> | 13 RHA | immediately behind 23 H |
> | Tac HQ | 069555 |
> | Main Bde | 050515 |

By around 0700 hours, 8 RB on the slower left route had caught up and was waiting to be deployed. Sergeant Hicks of H Company recalled that:

> By dawn we reached the outskirts of Amiens, which we could see quite clearly across the fields about a mile away ... with hedgerows criss-crossing the landscape. We pulled off the road into an open grassed field. The rain had stopped and the clouds began to break up and let the early morning sun

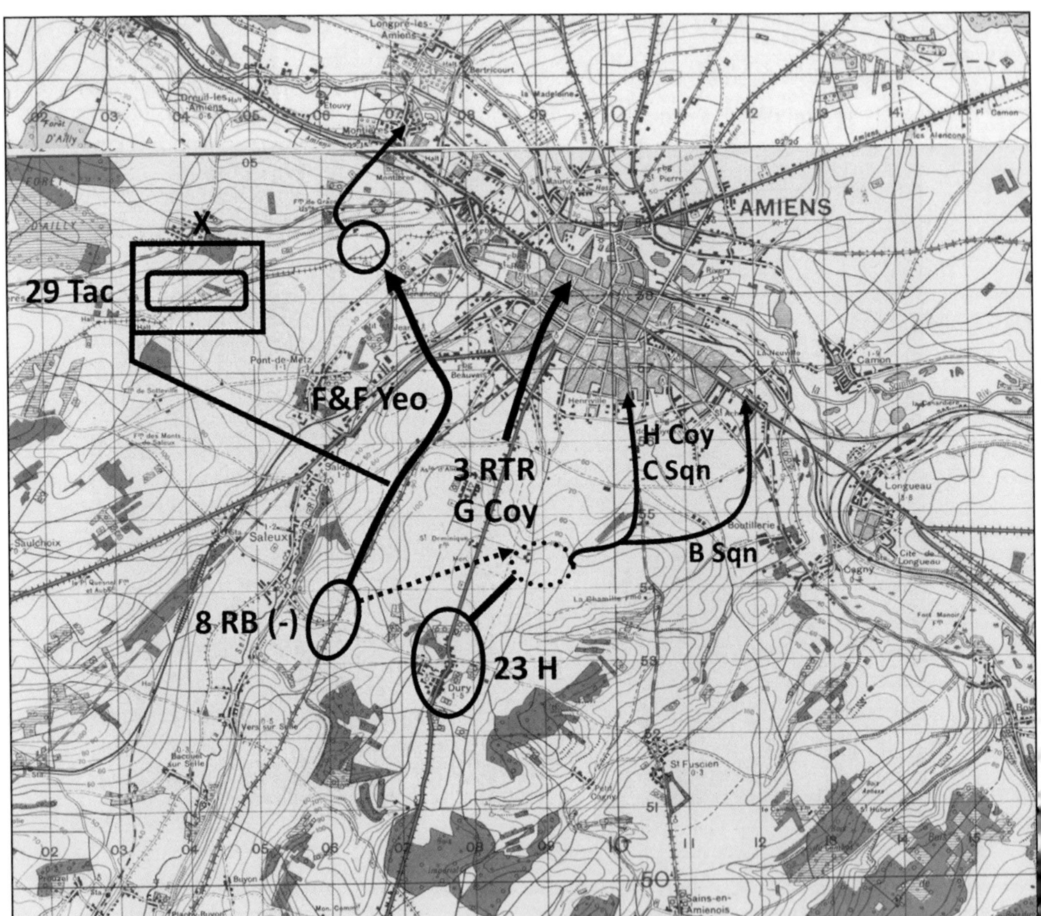

General Roberts' plan for the capture of the Somme bridges in Amiens on the morning of 31 August 1944.

filter through as we climbed out of our vehicles feeling very tired and bleary-eyed to stretch aching limbs, have a pee, and put the petrol cookers on for a brew up and cook something warm for breakfast.

The speed of the advance had clearly caught the Germans by surprise, so much so that the headquarters vehicles of the Seventh Army were captured by 23 H, along with maps showing the planned deployment of German formations on the line of the River Somme and to the south in detail. A little later, the army commander, General Eberbach,[1] himself fell into the hands of the Recce Troop of the F&F Yeo and the riflemen of F Company. It is recorded that he was having breakfast in a car parked on the road and according to General Horrocks he was still in his pyjamas. As a major prize, he was quickly sent back to divisional headquarters:

When they captured the German General Eberbach, we had to take him back to brigade headquarters. He was a very surly sort of gentleman. We had him right on the back of the scout car, although I think he had his legs inside. I saw three Germans running across the road into the woods. I said to this

General der *Panzertruppe* Heinz Eberbach.

German general … 'Tell those men, to come out here and get on the back of this scout car.' So, he did – he spoke English in a broken fashion. These men came out and jumped on to the back – and their eyes popped out of their head when they saw the German general! We went across this railway line and I saw a German walking along with his helmet in his hands. We stopped him. I got out of the scout car and had a look. In his helmet he had half a dozen eggs – which I said were mine and told him to jump on the back of the scout car. We delivered the German general to brigade headquarters – he didn't say goodbye!

When he reached divisional headquarters, it transpired that neither generals Roberts nor Horrocks were interested in speaking to Eberbach, 'they were simply too busy with the battle'. General Horrocks apparently asked what 'were all those Germans doing there, and would we please get them out of the way as soon as possible?'

Into the City

Not wasting any time, Colonel Silvertop of 3 RTR dispatched his A and C squadrons into Amiens with G Company but only, as recorded by Major Bell:

After a certain amount of discussion, negotiation, and I might even add, argument as to who should go first, the leading section of carriers and the leading troop of tanks arrived at a compromise by which they advanced abreast.

Major Bill Close, A Squadron's commander recalled:

At this stage my squadron consisted of ten tanks. Noel Bell, who came up to confer, told me rather plaintively that his company strength was seventy riflemen. Nevertheless, we decided that as the rest of 3RTR and the Rifle Brigade were coming up we would push on, the lead tanks being covered by the others.

As was the case with the advance from the Seine, the capture of Amiens, despite considerable confusion in German ranks, was not without opposition. Major Bell continued that: 'It was now getting really light and we could see enemy columns and single vehicles streaming into the city along roads running parallel to our own, and a certain amount of confused shooting took place.' Encounters with the enemy were in fact numerous both on the way into and once in the city, with 3 RTR noting that: 'The countryside was littered with burning wrecks, alive with running figures.' However, arriving in the outskirts of Amiens, G Company was informed by the resistance

that there were approximately five thousand Germans in Amiens, and at that early hour and having had no sleep, this information shook us a bit. Still, orders were orders, and we decided to push straight on, keeping to the main streets and making no diversions in case we got split up into small parties. If we were bold and quick we reckoned we could probably get to our objective before the Germans realised we were in the city, because it was quite obvious that up to the present they had not the slightest inkling of the presence of a British armoured column within miles of them.

Major Bell earned his Military Cross for 'acting in the boldest manner' as he and his company 'forced their way through the town'. Major Close recalled one of the obstacles that had to be forced:

Roadblocks are useless unless they are substantial and well defended. The enemy had been given no time to organise himself and at the first barricade all eyes looked for the opposition.

Bang! The enemy gunner must have been nervous. His shot demolished a hardware shop and before another could be fired, the position had been smothered in .5 Browning fire. An abandoned lorry and a couple of carts were swept aside by a Sherman and we moved on. Impressed by this success, excited *maquisards* climbed onto the back of the tanks and pointed out the way forward and the known enemy positions.

The buildings got taller and the back of my neck grew pricklier but gradually, shooting now and then, we made our way to the river. And there it was. A large bridge. Deserted. I scanned the houses on the distant bank

Major Noel Bell MC.

through binoculars but saw little sign of activity. The people of Amiens were sensibly lying low. So was the enemy.

The advance into Amiens was, however, in vain, as the Germans had prepared the Pont Boulevard des Planets and the other three militarily significant bridges over the Somme for demolition, but G Company and A Squadron were almost successful.

Lieutenant Yate's No. 1 Troop leading A Squadron crossed the bridge successfully and the riflemen of 11 Platoon were on their feet and approaching it covered by the reassuring bulk of the Shermans of A Squadron's headquarters. Major Close recalled that

> we emerged from the shelter of a side street, increased speed and were climbing the approach road when, only a few yards away, the bridge was enveloped in smoke and collapsed noisily into the river. Reversing hurriedly into cover, we waited for the dust to clear.

The blowing of the bridge was recorded at 0800 hours in 3 RTR's war diary. Shortly afterwards at least one of the other bridges in the city was blown. This, of course, left the 11th Armoured Division without a bridge for sustained military traffic and with G Company now fighting with groups of Germans across the southern part of the city they were in no position to secure a bridge. Major Bell wrote:

> To cut a long story short, we entered Amiens, the fighting inside the city developing into a series of individual platoon actions, in which all achieved

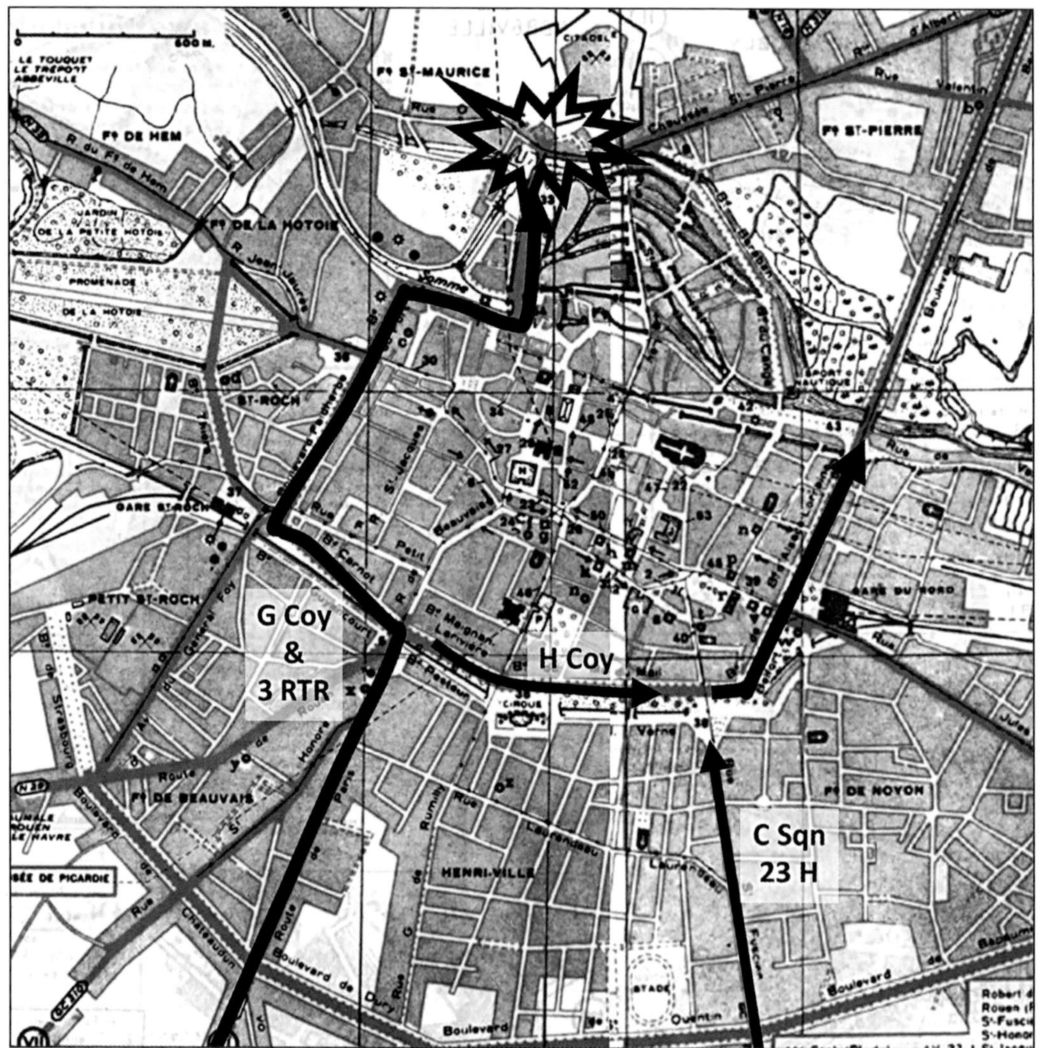

Operations to capture the Somme bridges.

notable success and the Germans' forces were knocked clean off their balance and mopped up before they really had any chance of realizing what was happening.

With 3 RTR and G Company committed, at 0830 hours, Brigadier Harvey ordered 23 H and the F&F Yeo, both deployed on the high ground south and south-west of Amiens, to recce east and west, respectively, for crossings. The tanks of 23 H's C Squadron and H Company would meet in the city and go for the Pont d'Amont, while B Squadron was to take bridges in the eastern suburbs. C Squadron's account reads:

The route to the city, almost a mile away, lay across open country. The map showed a track leading from Dury almost directly to our objective, but after we had moved down it over the crest of the hill, to within full view of the city, it petered out completely. A way, however, was found into the outskirts and soon we reached the built-up area, where excited civilians crowded round – one of them a schoolmistress who spoke good English – and told us that enemy tanks had pulled out to the east, but that there were still Germans in the town intending to blow the bridges.

With No. 1 Troop leading, the squadron entered the city via Rue St Fuscien, while H Company motored in from the south-west, rendezvousing with the tanks at Place du Maréchal Joffre. C Squadron's account continued:

The order now came to push on with all possible speed. Everywhere the clatter of tanks and carriers on the cobbled road brought excited spectators to the windows, hardly able to believe their eyes at this sudden and un-expected sight of British tanks right in the heart of their city and so very much on top of the Germans. After a moment's halt to make certain of directions, the leading tanks swung right into the tree-lined Boulevard d'Alsace Lorraine and made for the bridge.

Sergeant Triggs' carrier section led the way to the bridge, at best speed followed by the rest of the company-squadron group. They reached a point 100 yards from the bridge 'when they were brought to a sudden halt by fire from buildings on the opposite side'. The bridge was clearly prepared for demolition, with 'ominous

But for the dash of 11th Armoured Division's descent on Amiens, amidst substantial buildings the fighting could have been protracted.

black holes across the centre span'. Nonetheless, Sergeant Triggs led the way across:

> The RBs had dismounted and taken up fire positions on either side of the road. A crowd of Maquis, including women, armed with rifles and grenades had suddenly appeared. The leading tanks and carriers darted across, wildly followed by the Maquis, who rushed on in pursuit of the scattered German rearguard, leaving Fourth Troop to guard the river bridge.

Following the leading tanks, 14 Platoon crossed the bridge to complete a firm defence. On the northern side of the river bridge was an area of marshland that was crossed by an embankment/viaduct and a smaller bridge over a subsidiary course of the Somme. These, of course, had to be secured by the riflemen and tanks to gain a crossing. Along with the other motor platoons of H Company

> First Troop again went forward to secure the crossing of a viaduct vitally important to the advance, and again, with the assistance of the RBs and the Maquis, a small German counter-attack was driven off. Now the whole Squadron group moved across and established a tight, firm bridgehead.

To complete the reduction of the River Somme as a basis for German defence, the F&F Yeo and F Company secured a bridge at Montières, 1½ miles north-west of the city, that was well defended in terms of numbers but, lacking fight, a hundred prisoners were taken quickly. Further east, the armoured cars of 2nd Household Cavalry Regiment, advancing ahead of the Guards Armoured Division, captured a bridge at Corbie intact. Further right, the US Army was also across the river.

Fighting in the City

In the dash to reach the bridges quite a number of Germans had been rudely awakened by the eruption of British armour in the city and naturally wanted to escape:

> The populace soon began pouring out into the streets shouting and scream-ing with joy which initially made life very difficult for us because many groups of Germans were coming out of their barracks etc. and were trying to put up some show of resistance as well as sniping at us from some of the buildings. It was all very half-hearted on their part and the many skirmishes throughout the town soon petered out in our favour and I don't believe we suffered any casualties although there was some amongst the civilians and many Germans.

One reporter in a piece widely syndicated in regional newspapers described the action of Lieutenant Sudlow's 10 Platoon of G Company:

> He was in command of the leading platoon of a battalion of the 11th Arm-oured Division, which entered Amiens side by side with the tanks. On the

A pre-war postcard showing the entrance to Gribeauval Barracks in Amiens.

platoon's way down the main street it was noticed that a big Hitler barracks had received the attention of our bombers. A road ran alongside the barracks and at the far end connected up with another parallel main road into the city.

The latter appeared to be popular with retreating Germans, so Lieut. Sudlow put out three Bren gun teams behind piles of rubble and covered the far end of the connecting road, much in the style of a rifle range.

The first target was a lorry, which came tearing round the corner and was stopped dead with simultaneous bursts from the Brens. Then a section of infantry tried to double across the opening. Of the dozen who tried only two succeeded. Another lorry gasped to a standstill and some more infantry fared even worse, for by now the marksmen were getting warmed up and competition was pretty keen.

A staff car almost made the crossing, but was stopped by a brilliant shot, which killed the driver. The passenger jumped out and dived for cover under one of the lorries. The lorry was shot up. A large number of infantry tried to silence the post but failed miserably. Fifty prisoners were taken and as many left wounded or dead. The scene at the far end of the road was indescribable.

The fighting was short lived, but prisoners were being rounded up throughout the day. For example, 4 KSLI had been well forward in the column with the expectation that they would be committed to some serious fighting in the city but in the event when they entered the city at 1400 hours, according to their war diary:

There was practically no fighting at all as the Bn pushed forward and took up position. We had a tremendous reception, which was almost embarrassing in its intensity. Large numbers of prisoners were taken in co-operation with the F.F.I. [French Forces of the Interior], whose help was invaluable.[2]

Fighting Back on the Route to Amiens

The night move to Amiens that had so successfully caught the Germans by surprise had, of course, punched through numerous enemy units who were now bent on getting east to the Somme. Consequently, on both of the routes to Amiens, the crews of various 8 RB carriers and half-tracks that had either broken down or run out of fuel were having serious contacts with parties of enemy who were in many cases still full of fight. One such incident was recorded by Sergeant Fruin, whose carrier ran out of fuel at about 0400 hours during the march to Amiens:

> We stopped several vehicles which came along, but the answer was always the same, 'Sorry, chaps, nearly out ourselves,' so we decided the only thing

The long logistic tail of the division, which would include 8 RB's B Echelon, crossing DAVID at Vernon. With a change in the weather the sappers are wearing gas capes to keep dry.

to do was to wait for the Echelon, which we knew would be following on behind.

By this time the sun was just coming up and we found ourselves on a stretch of road which was lined with trees. A good ditch on the left-hand side we decided would be an excellent place for a brew and a spot of breakfast, which was well under way when Cripps spotted the Echelon emerging from a wood on what appeared to be the road on which we had halted. The very misty dawn, however, had deceived us, and imagine our surprise when we discovered that the Echelon was a German one moving along a side road, joining our road about 50 yards behind us.

The enemy column of eleven trucks and a Kubelwagen was about 600 yards from them when the riflemen opened fire with a Bren gun and three rifles, attempting to halt them at a respectable distance. This halted a vehicle in the centre of the column but the ones in front continued, turning on to the main road:

We simply could not miss the leading vehicle and, letting the driver's cab have a burst, it came to a standstill amid clouds of steam, only 30 yards from us. Two Germans leapt out of the back and dived for the ditch, but that was the last thing they ever did.

The remainder of the column panicked and drove off the road in all directions, to no avail, however, as within the next few seconds we had 'brewed up' four more vehicles and the others just stopped, their crews baling out. The next half-hour was spent in playing hide-and-seek in the corn, exchanging shot for shot, and we collected in twenty-three prisoners, counting in addition five dead and nine wounded. Our own casualties consisted of one spoilt brew.

One of the prisoners spoke a little English, so we told him to go and get some petrol from his vehicles. This he seemed only too pleased to do and, having filled up, we continued our journey to rejoin the rest of the Company.

The Somme Defences

Capturing the bridges in Amiens played a vital part in thwarting German plans to stabilise a front on the River Somme but it would seem that there was one part of the German defences that was properly constituted. This was a screen on the rising ground north of the city, based on 105mm or 88mm flak guns and sundry smaller weapons, whose crews had plenty of time to man their weapon pits. By the time the bridges had been taken, these guns were primarily deployed in the ground role positioned to cover the exits from the city and other crossings to the east and west.

By 1130 hours the leading battalion of 159 Brigade was entering Amiens and at 1340 hours, with all the individual battles resolved, the 29th's war diary recorded: 'Bde ordered to pass through AMIENS and est itself on the high ground NE of the town.' From their bridgeheads, 23 H and F&F Yeo were to lead with their motor companies under command, while 3 RTR followed as brigade reserve.

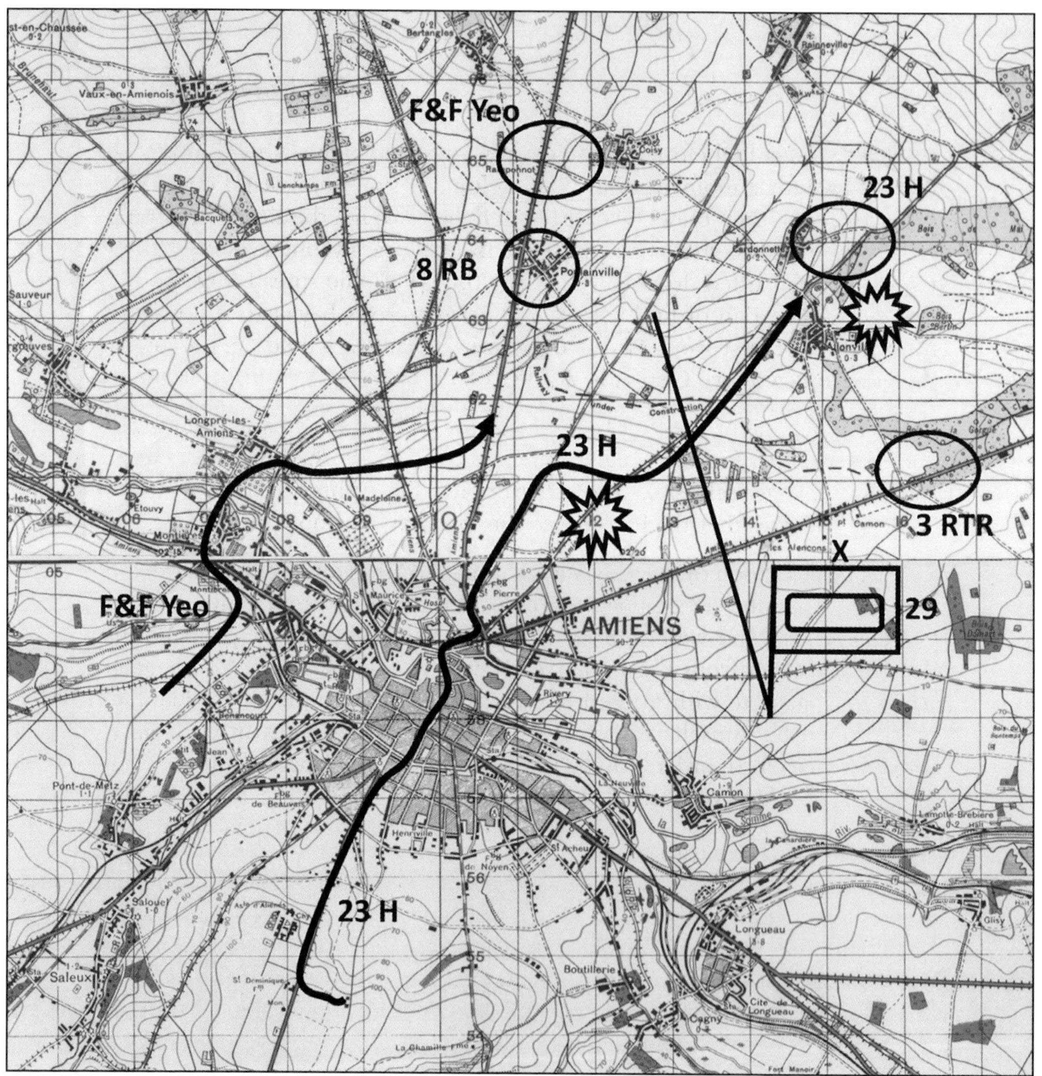

The advance across the Somme during the afternoon of 31 August. Also shown is the approximate locations of the two batteries encountered by H Company and 23 Hussars.

The Hussars led across the main bridge into Amiens' northern suburbs:

Then a piece of great good fortune occurred. The road out of the town, which 'B' Squadron should have used, was covered by a large Flak site. Fortunately a slight error of map-reading was made and they did not use that road. Instead, they took a smaller one which brought them out in open country in the rear of the Flak site, which still had its guns pointing down the road they should have taken. After one incredulous glance, 'B' Squadron opened up with all they had, assisted by RHQ and the leading troop of 'C' Squadron. The defenders were completely surprised – they had at first supposed that our tanks were German. As we opened fire, frantic activity began. Large guns began slowly to swing towards us, but each one in turn received a direct hit before it could get any tank in its sights. One by one they disappeared for a moment in a sudden cloud of grey smoke and dust. The black figures of their crews could be seen flying through the air, to land some yards away and lie still. The big, long barrels one by one ceased turning towards us and tilted drunkenly towards the sky. Panic spread through the Flak site as the rain of fire increased. The gun crews began to run blindly towards the nearest wood, while every minute a shell would land amongst them and a few more Germans would fall.

Captain Bailey, one of the FOOs of 64 Medium Regiment's battery that was in support of the 11th Armoured Division and moving with the F&F Yeo, engaged the enemy anti-aircraft site. The citation for his Military Cross earned on 31 August indicates that the battle with the flak guns was not entirely one sided:

In the late afternoon of that day, he was informed that four 88mm guns were engaging a number of our tanks (which were on a different centre line to Captain Bailey) and were causing casualties amongst them. Captain Bailey detached himself from the squadron to which he was attached and man-oeuvred his tank to within about 600 yards to the rear of the enemy battery: in so doing, he himself was engaged by two other 88 mms firing from a different quarter. In order to improve his observation, Captain Bailey left his tank and crawled considerably nearer to the four 88s using his 38 set as

An 88mm Flak 36 and crew training.

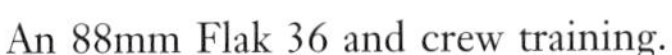

means of communication. During all this time, he was under continual fire from the other two 88s, which were mostly using airburst [HE]. Having found the best possible place from which to observe the enemy battery, Captain Bailey brought the fire of his regiment to bear and in a short time had silenced all four guns, observing many near misses on each gun and seeing the gun crews running for cover. Throughout the shoot, this officer was in danger of being hit by splinters from his own shelling owing to the proximity of the target. Through his total disregard for his own safety, casualties to our own tanks were stopped and they were able to continue their advance.[3]

When Captain Bailey had adjusted the fall of the shells, five round 'fire for effect' from each of the eight 5.5-inch guns of the medium battery was more than enough to neutralise the enemy guns.

A troop of tanks advanced on the enemy position and: 'All around the tanks shaken figures in field grey rose to their feet, hands above their heads.' No doubt the riflemen of H Company helped round up the prisoners and send them to the rear.

With just 2 miles further to their overnight harbour location, the leading tanks of the Hussars' A Squadron were approaching Allonville when they came across another German position, this time consisting of fifteen light anti-aircraft guns. H Company noted: 'In the afternoon we moved out to ACTONVILLE where, after shooting at some completely disorganised Germans, we settled down for the night.'

Sergeant Hicks summed up his feelings at the end of a remarkable two days of 'swanning' that had taken 8 RB out of Normandy and well on the way to Belgium:

> By the afternoon the whole task force had consolidated a bridgehead on the other side of the River Somme and to put it mildly I think we were all feeling rather knackered by all the excitement of the last couple of days coupled with the lack of sleep. At the same time, we were quietly pleased with all our achievements.

As 1 September dawned, Antwerp, which had seemed to be 'a very distant objective', now seemed entirely plausible. However, those who believed that the 1918 collapse of the German Army was being repeated, including a SHAFE staff officer who wrote 'end of the war is within sight, almost in touch', were to be disappointed. General Eisenhower's broad front strategy proved to be logistically unsupportable and the Allies' advance ground to a halt at the end of the first week in September with the armies being far from 'lined up on the Rhine'.[4] Consequently, there would be another bitter winter at war and for the riflemen, another seven months of costly fighting. Despite this, having advanced 340 miles from the Seine to Antwerp in just six days, 8th Rifle Brigade, along with the rest of the 11th Armoured Division, had entered a new phase of the North-West European Campaign, which is altogether another story.

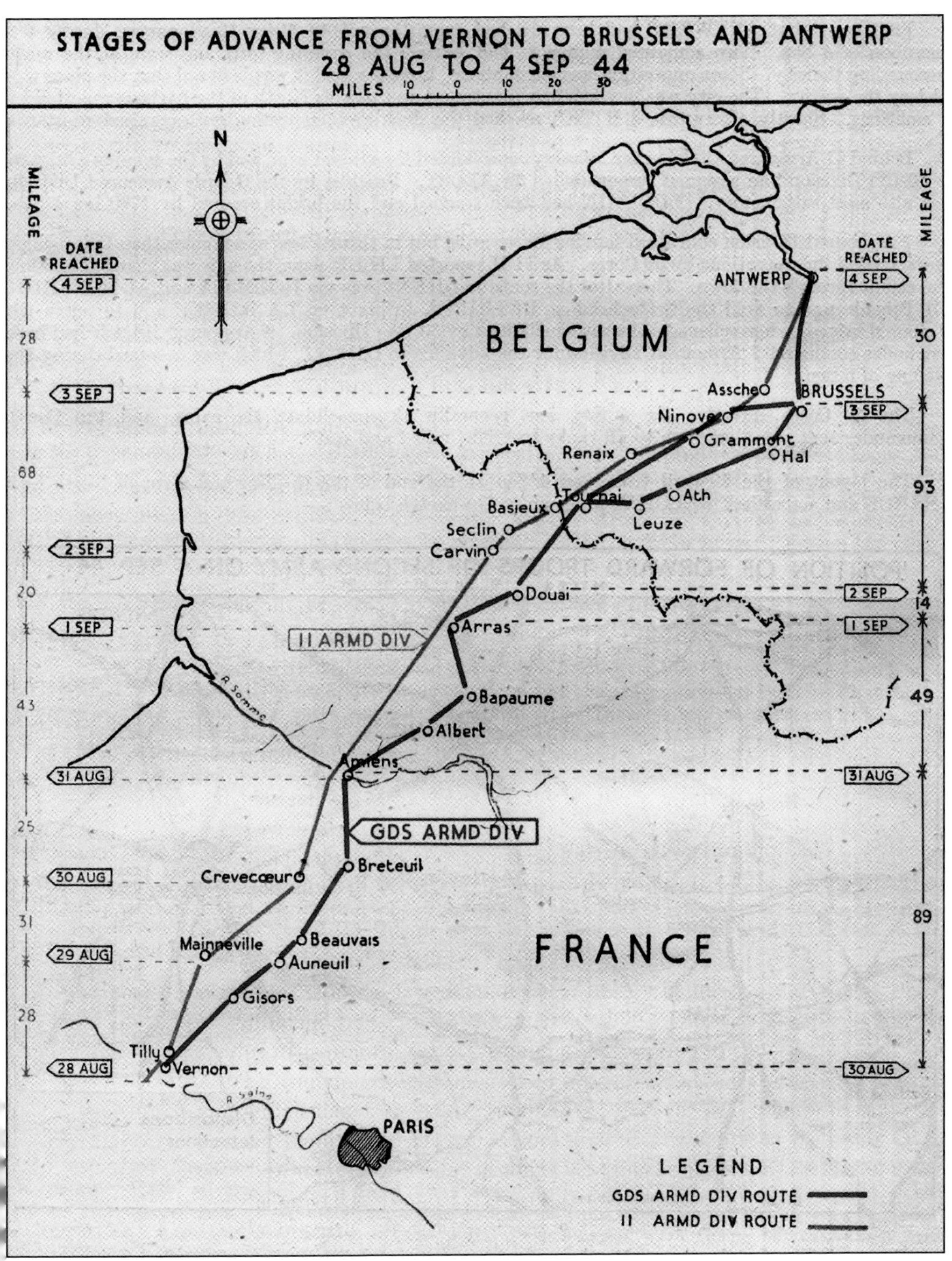

The Great Swan across northern France and into Belgium by the 11th and Guards armoured divisions.[4]

The rifleman's small pack and its contents, which along with his basic webbing, completed his fighting order. (*Mathew's Military Moments*)

Notes

Introduction

1. The units and subunits of 11th Armoured Division were encouraged to write their own histories. Beyond this, the sabre squadrons of 141 (The Buffs) Regiment RAC are the only other subunits known to the author to have written accounts.

Chapter One: In the Beginning

1. Hasting, Major RHWS, *The Rifle Brigade in the Second World War 1939–1945* (Naval & Naval Press Ltd, Uckfield).
2. Anon, *Taurus Pursuant, A History of 11th Armoured Division* (Printing and Stationary service, British Army of the Rhine, 1945).
3. This was 30 Armoured Brigade, which included 12th KRRC as its motor battalion.
4. CIGS, *MTP No. 41, The tactical handling of the Armoured Division and its components, Part 3: The Motor Battalion* (War Office, June 1943).
5. This inadequacy had been highlighted in the reports of the Experimental Mechanised Force in the early 1930s but producing a tracked vehicle for the infantry was at the time unaffordable.
6. Anon, *Taurus Pursuant.*
7. There was a particular emphasis on high-profile visits to the east coast and the north in support of Operation FORTITUDE NORTH and the fictitious Fourth British Army.
8. BBC Peoples' War.
9. The heavy equipment and wheeled vehicles of the division sailed from Tilbury, while the infantry battalions crossed the Channel in LCIs from the south coast port of Newhaven and the tanks were loaded onto LCTs at hards in the Solent.

Chapter Two: Normandy at Last

1. Annon, *The Story of the Twenty-Third Hussars 1940–1946* (Naval and Military Press Ltd).
2. *Administrative History of 21 Army Group 6 June–8 May 1945* (HQ BAOR, November 1945).
3. TNA WO 171/1359.
4. TNA WO 171/866.
5. Jefferson, Roland, *Soldiering at the Sharp End* (unpublished journal).
6. Tac HQ 21 Army Group directive M504.

Chapter Three: Operation EPSOM

1. Roberts, Pip, *From the Desert to the Baltic* (William Kimber, London, 1987).
2. The village of Cheux (*Sher*) was referred to by everyone as 'Chucks' or 'Chooks'.
3. TNA WO 171/456.
4. Conversation Hubert Meyer and author, 2005.
5. TNA WO 171/853.
6. Air Historical Branch, trans. 1948, *Telephone Log of the German 7th Army, June 6–June 30, 1944.*
7. Anon, *The Story of the 23rd Hussars* (Naval and Military Press, Uckfield).

Chapter Four: The First Battle of Hill 112

1. Anon, *Operations of Eighth Corps – Normandy to the Rhine* (St Clements Press Ltd, London, 1948).
2. BBC People's War.

3. Anon, *The Story of the 23rd Hussars* (Naval & Military Press, Uckfield).
4. The location and orientation of Hill 112's woods are correct on the 1:25,000 map Cheux, sheet 37/16 S.E.
5. Meyer, Hubert, *The History of the 12. SS-panzerdivision Hitlerjugend* (Fedorowicz, Canada, 1994).
6. A part of III Flak Corps, most of which was deployed against the Second Army.
7. Oxford Reference, 'Ground of such importance that it must be retained or controlled for the success of the mission'.
8. A YOKE target was one of the concentrations of all guns available and in range that could be called by a FOO: MIKE – regimental, YOKE – divisional, UNCLE – corps, VICTOR – army. In one such engagement the Ayrshire Yeo fired 140 rounds per gun.
9. TNA WO 171/627.
10. II SS Pz Corps' commander, *Obergruppenführer* Hausser, explained during his post-war interrogation that: 'It was scheduled to begin at 7 o'clock in the morning, but hardly had the tanks assembled when they were attacked by fighter-bombers. This disrupted the troops so much that the attack did not start again till two-thirty in the afternoon.'
11. AHB paper, *War Time Experience – 1940 to 1945, Rocket Projectiles* (undated).
12. Jackson Lt Col GS, *Operations of Eighth Corps – An Account of Operations from Normandy to the Rhine* (St Clement's Press Ltd, London, 1948).

Chapter Five: Between Battles

1. This was the result of fighting between 26 SS Panzer Grenadiers and the 3rd Canadian Division on 8–9 June 1944.
2. 101 Reinforcement Group to the immediate east of Bayeux, south of the Crépon road.
3. Hicks, Mike, *Difficult Days: A Rifleman's Wartime Memoir 1939–1946* (Indepenpress Publishing Ltd, 2014).
4. Lt Elgood, the anti-tank platoon commander that 'abandoned' the three surviving guns of G and H companies in the wood during the withdrawal, came in for a lot of blame, but the battalion recovered the guns before the Germans reoccupied the hill.
5. TNA WO 171/866.
6. So heavy were casualties in battle that a new planning measure was introduced, 'Double Intense'.
7. This was one of the divisions formed from surplus Luftwaffe manpower. They never had a good reputation and were taken into the Army.
8. The Reich Minister for Armaments and War Production ordered the formation of Baukommando Becker. It was based in three French factories; the Matford, the Talbot and the Hotchkiss factories around Paris but had strong links to firms in Germany.
9. *Schnelle* brigades were conceived as highly mobile formations based in France as a rapid reaction force. Being entirely motorised, they were intended to cover distance quickly, to reach the scene of an invasion.
10. Feuchtinger was an artillery officer and Nazi party organiser of events such as the Nuremburg Rallies. He is widely believed to have gained his appointment on the back of these organisational skills and party interest rather than distinguished military service.
11. TNA CAB 196/959.
12. Three additional bridges were to be built, construction starting at GOODWOOD's H Hour. They would, however, not open until the evening of the 18th at the earliest.
13. Unlike the other bridges, TOWER Bridge had a nickname for GOODWOOD: YORK.

Chapter Six: Operation GOODWOOD

1. Close, Major Bill, *A View from a Turret* (Delle & Bredon, Tewkesbury, 2002).
2. Lehmann and Tielmann, *The Leibstandarte IV/1* (J.J. Fedorowicz, Canada, 1993).
3. Bell, Major Noel, *From the Beaches to the Baltic, G Company 8th Rifle Brigade* (Privately published, Germany, 1946).
4. TNA WO 171/853.

5. Anon, *Taurus Pursuant* (privately published 1945).
6. TNA WO 171/1359.
7. Author's interview with Major Bill Close, 2004.
8. Lt Noakes, a liaison officer with the F&F Yeo, noted that: 'Scout cars are not good over rough country, ditches and banks – and they are dreadful over railway lines.'
9. IWM Sound Archive AC 20318.
10. Quoted in Hart, *Burning Steel*.
11. TNA WO 171/627.
12. Staff College battlefield study 1956.
13. Hicks, Mike, *Difficult Days: A Rifleman's Wartime Memoir 1939–1946* (Indepenpress Publishing Ltd, 2014).
14. Directorate of Army Training, Operation Goodwood, (SSVC video, 1975).
15. Hastings, Major HWS, *The Rifle Brigade in the Second World War 1939–1945* (Naval & Military Press reprint).

Chapter Seven: Operation GOODWOOD – Day Two

1. TNA WO 171/1359.
2. Hicks, Mike, *Difficult Days: A Rifleman's Wartime Memoir 1939–1946* (Indepenpress Publishing Ltd, 2014).
3. Lehmann and Tielmann, *The Leibstandarte IV/1* (J.J. Fedorowicz, Canada, 1993).
4. Jackson, Lieutenant Colonel, *Operations Eighth Corps Normandy to the River Rhine* (St Clements Press Ltd, London, 1948).
5. TNA WO 171/997.
6. Anon, *The 1st and 2nd Northamptonshire Yeomanry* (Joh Heinr Meyer, Germany).
7. *The Tactical Handling of the Armoured Division and its Components, Military Training Pamphlet No. 41, Part 3 The Motor Battalion.*
8. Lt Sedgwick was one of those wounded during street fighting training in the blitzed part of Limehouse during January 1944. He had been shot in the nose by a round from a Sten gun. He was killed in action later in the campaign.
9. Hastings, Major RHWS, *The Rifle Brigade in the Second World War 1939–1945* (Naval and Military Reprint).
10. Jackson, Lieutenant Colonel, *Operations Eighth Corps Normandy to the River Rhine* (St Clements Press Ltd, London, 1948).

Chapter Eight: A Move West

1. I Corps' rest centre was closest to Cussy, being on the sea front at Lion-sur-Mer.
2. TNA WO 171/1359.
3. At the same time the armoured regiments all received the best part of a squadron from the disbanded Royal Gloucester Hussars.
4. TNA CAB 106/963 *Notes on the fighting in Normandy – Report No. 12* (17 June 1944).
5. 21 A GP 2182/Ops/1/(B). Enclosed in CAB 106/963.
6. TNA WO 171/456.
7. Anon, *Taurus Pursuant, A History of the 11th Armoured Division* (privately published 1945). This is a direct quote from the divisional war diary on 29 July (TNA WO 171/456).
8. The author as a junior army officer developed a taste for Calvados while negotiating with Norman farmers for access to their land for battlefield studies.
9. *Administrative History of 21 Army Group* (Germany, November 1945).

Chapter Nine: Operation BLUECOAT

1. Anon, *The Story of the 23rd Hussars* (N and M Press, Uckfield).
2. Roberts, Major General GPB, *From the Desert to the Baltic* (William Kimber, London, 1987).
3. Steel Brownlie, *The Proud Trooper* (Collins, London, 1964).

4. 'Standing to Arms': the British Army practice was for all ranks to be in their battle positions, fully accoutred for an hour astride dawn and dusk, favourite times for enemy attack. Stand-to also marked the transition between day and night routine.

5. In one of the numerous errors in copying the French 1:50,000 map, Charles de Percy is wrongly named. What is named is La Ferronnière on the Vier road (COVENTRY), the real Charles de Percy is a mile further on. British cartographer's practice is to put place names to the right of a feature, while the French map makers generally place them where convenient. A source of much confusion!

6. The Inns of Court Yeomanry had replaced 2 HCR as the division's armoured car attachment the day before.

Chapter Ten: Le Bas Perrier

1. Hicks, Mike, *Difficult Days: A Rifleman's Wartime Memoir 1939–1946* (Indepenpress Publishing Ltd, 2014).
2. Jefferson, Rolland, *Soldiering at the Sharp End*, unpublished journal.
3. McCully, Sergeant Bertie, unpublished memoir.
4. The A1 Echelon was normally several miles behind the Fighting Echelon, ready to provide immediate resupply of ammunition and fuel. It was also where the company and squadron fitters were located.
5. BBC People's War.
6. TNA WO 171/627.
7. Robert, Major General CBF, *From the Desert to the Baltic* (William Kimber, London, 1987).
8. In battle corps and divisions were usually reinforced by RASC ambulance companies.

Chapter Eleven: Pursuit to the Seine

1. 9th Armoured Division had been planned to be one of the formations that would take part in OVERLORD but had been replaced in the ORBAT by 7th Armoured Division on its return from Italy, as Montgomery wanted experienced divisions to spearhead the campaign. It was disbanded on 31 July and its units either broken up to provide individuals reinforcement or sub-units. The Royal Gloucester Hussars are an example of the former and 8 KRRC the latter. Others such as the 15th/19th Hussars were deployed to replace understrength units, in their case replacing 2 N Yeo in the 11th Armoured Division. Regular Army units were always favoured over TA or war-raised units when it came to disbandment and/or deployment.
2. TNA WO 171/627.
3. Anon, *The Story of the 23rd Hussars* (N&M Press, Ltd).
4. Anon, *Taurus Pursuant: A History of the 11th Armoured Division* (Privately published, 1945).
5. Some accounts mention 9th *Hohenstaufen* SS Panzer Division, but they were not in the same area and they could have mistaken the camouflage clothing of 21st Panzer for SS. However, in the confusion of the Falaise Pocket anything was possible.
6. The F Company history describes this aircraft as a 'flying Jeep'.
7. Noted in the regiment's war diary: 'In little more than 5 days therefore, the Regt with no previous warning, had been moved overseas, re-equipped with entirely fresh equipment and taken over a new role in a fresh formation and taken its place in the battle, which it had for so long been wishing to do. The difficulties of all this very rapid movement and equipping and of taking on a new role amongst strange people inevitably had its effect and for some time it is doubtful if the Regt performed as well as it was capable of doing.'
8. Roland Jefferson, *Soldiering at the Sharpe End* (Unpublished memoir).
9. XXX Corps' axis/main supply route nicknames were CLUB, HEART, SPADE and DIAMOND.

Chapter 12: The Great Swan

1. Simpson's Ninth US Army was added to the order of battle in mid-September.
2. Horrocks, Sir Brian, *Corps Commander* (Sidgwick and Jackson, 1997).

3. Hicks, Mike, *Difficult Days: A Rifleman's Wartime Memoir 1939–1946* (Indepenpress Publishing Ltd, 2014.

4. These classes refer to the carrying capacity of the bridges for vehicles in tons.

5. TNA WO 171/627.

6. It would seem that on this occasion 13 RHA were supporting the tanks and the 3-inch mortars of F and G companies the dismounted attack on Étrépagny.

7. Having outrun the available 1:50,000 mapping, XXX Corps was working off 1:10,000 maps, which were notoriously inaccurate in the detail required by the fighting troops at the front.

8. 29 Armd Bde Op O No. 15, signed 0100 hours 30 August 1944.

9. Delaforce, Patrick, *The Black Bull – From Normandy to the Baltic with the 11th Armoured Division* (Alan Sutton, Stroud, 1993).

10. Roberts, Major General GPR, *From the Desert to the Baltic* (William Kimber, London, 1987). The brigade war diary makes it clear that 8 RB were behind the F&F Yeo not in the main column.

11. IWM sound archive Cat No. 19867.

12. IWM sound archive Cat No. 20315.

Chapter 13: The Amiens Bridge

1. General Eberbach took command of the 7th Army when his predecessor General Hauser was wounded in the Falaise Pocket and was attempting to mount a defence on the line of the Somme.

2. TNA WO 171/1326.

3. TNA WO 373/51/145, folio 258–259.

4. *An Account of the Second Army in Europe, Vol. 1* (HQ Second Army, Germany 1945).

5. Some would add as a factor here General Marshal's desire to use the First Allied Airborne Army causing Eisenhower to halt XXX Corps' and Patton's drives to the Rhine in favour of the broad front and MARKET GARDEN's advance to Arnhem.

Panthers and German transport knocked out in the Falaise Pocket.

Index